Internet of Things - New Insights

Edited by Maki K. Habib

Published in London, United Kingdom

Internet of Things - New Insights
http://dx.doi.org/10.5772/intechopen.111033
Edited by Maki K. Habib

Contributors
Mohammed M Alenazi, Magauwane Reneilwe Maepa, Michael Nthabiseng Moeti, Zahra'a Mohammed Baqir, Hassan J. Motlak, Francisco A. Delgado Rajó, Ione Adexe Alvarado Ramírez, Cajetan M. Akujuobi, Faith Nwokoma, Xishuang Dong, Lucy Nwosu, Sheikh Rufsan Reza, Xiangfang Li, Călin Corciovă, Robert Fuior, Alexandru Sălceanu, Cătălina Luca, Yu Chen, Qian Qu, Han Sun, Vesna Antoska Knights, Zoran Gacovski, Maged E. A. Mohammed, Nashi K. Alqahtani, Muhammad Munir, Mohamed A. A. Eltawil

First published in London, United Kingdom, 2024 by IntechOpen
IntechOpen is the global imprint of INTECHOPEN LIMITED, registered in England and Wales, registration number: 11086078, 5 Princes Gate Court, London, SW7 2QJ, United Kingdom

British Library Cataloguing-in-Publication Data
A catalogue record for this book is available from the British Library

Additional hard and PDF copies can be obtained from orders@intechopen.com

Internet of Things - New Insights
Edited by Maki K. Habib
p. cm.
Print ISBN 978-1-83768-987-3
Online ISBN 978-1-83768-988-0
eBook (PDF) ISBN 978-1-83768-989-7

For EU product safety concerns:
IN TECH d.o.o., Prolaz Marije Krucifikse Kozulić 3, 51000 Rijeka, Croatia,
info@intechopen.com or visit our website at intechopen.com.

Meet the editor

Professor Maki Habib's distinguished career in control, computer systems, mechatronics, and robotics reflects an unwavering commitment to innovation. He earned his BSc and MSc in Control and Systems Engineering from the University of Technology - Baghdad, Iraq, and a Doctor of Engineering Sciences in Intelligent and Autonomous Robotics from the University of Tsukuba, Japan. Throughout his distinguished career, Professor Habib held vital roles at renowned institutions globally, from research scientist at RIKEN, Japan, to senior researcher at RISO-Laboratories, Japan, and later as an associate professor at University of Technology Malaysia (UTM). He demonstrated exceptional leadership as a senior researcher at GMD-Japan and played a leading role in developing the program in Mechatronics Engineering at Monash University, Australia. Professor Habib's impact extends beyond academia to consulting for industrial giants like Toyota, Denso, and ABB. His influential collaborations in European research consortia significantly contributed to advancements in robotics, Internet of Things (IoT), Cyber-Physical Systems (CPS), and sensors, with projects ranging from advanced robots for demining to portable IoT kits supporting sustainable agriculture. As an esteemed scholar, Professor Habib served in editorial capacities for numerous international journals and contributed prolifically to literature with more than 24 books, 38 book chapters, 90 journal papers, and 245 conference papers. His diverse research interests, from nanorobotics to machine learning in cancer detection, showcase his dedication to impactful advancements. Professor Habib's excellence gained global recognition, placing him among the top 2% of the most impactful scientists worldwide annually since 2020. His career is a testament to his dedication and contribution to the dynamic fields of robotics and mechatronics, positioning him as a leading figure in these disciplines.

Contents

Preface IX

Chapter 1 1
A Systematic Review on IoT-Based Smart Technologies for Seat Occupancy and Reservation Needs in Smart Libraries at Institution of Higher Learning
by Magauwane Reneilwe Maepa and Michael Nthabiseng Moeti

Chapter 2 31
Methods for Detection and Prevention of Vulnerabilities in the IoT (Internet of Things) Systems
by Vesna Antoska Knights and Zoran Gacovski

Chapter 3 49
Applications of AI and IoT for Advancing Date Palm Cultivation in Saudi Arabia
by Maged Mohammed, Nashi K. Alqahtani, Muhammad Munir and Mohamed A. Eltawil

Chapter 4 87
Application of Internet of Things (IoT) in Biomedicine: Challenges and Future Directions
by Robert Fuior, Alexandru Sălceanu, Cătălina Luca and Călin Corciovă

Chapter 5 109
A Survey on IoT Fog Resource Monetization and Deployment Models
by Cajetan M. Akujuobi and Faith Nwokoma

Chapter 6 127
Design and Implementation Smart Petrol Station Using Internet of Things
by Zahra'a M. Baqir and Hassan J. Motlak

Chapter 7 167
Hybrid Architectures to Improve Coverage in Remote Areas and Incorporate Long-Range LPWAN Multi-Hop IoT Strategies
by Francisco A. Delgado Rajó and Ione Adexe Alvarado Ramírez

Chapter 8 185
IoT and Energy
by Mohammed M. Alenazi

Chapter 9 205
Effective Screening and Face Mask Detection for COVID Spread Mitigation Using Deep Learning and Edge Devices
by Xishuang Dong, Lucy Nwosu, Sheikh Rufsan Reza and Xiangfang Li

Chapter 10 231
Smart Healthcare at Home in the Era of IoMT
by Qian Qu, Han Sun and Yu Chen

Preface

Welcome to a new era in technology and connectivity where the Internet of Things (IoT) transforms our interactions with the world. *Internet of Things – New Insights* explores this dynamic and complex field, covering its technological evolution, diverse applications, key features, and synergy with other emerging technologies like Cyber-Physical Systems (CPS).

The journey of IoT is a fascinating exploration of innovation, tracing its evolution from early ubiquitous computing concepts to today's interconnected ecosystem. This book highlights the milestones and emphasizes the importance of integrating IoT, investigating the interaction between physical and digital processes, enhancing capabilities, and propelling advancements in automation, control, and decision-making.

As we embrace the boundless possibilities of IoT, it is also necessary to confront the challenges accompanying its rapid developments, deployments, and contributions to domains. Security concerns, interoperability issues, and ethical considerations are among the hurdles that demand thoughtful examination. The chapters in this book engage with these challenges, offering insightful perspectives and potential solutions. This helps to set the stage for a secure and resilient IoT landscape, paving the way for the transformative future discussed in the book.

This book is designed for a diverse audience, including researchers, professionals, curious enthusiasts, and students eager to understand the complexities of IoT. Whether you are an industry expert seeking in-depth insights or a student looking to grasp the fundamentals, the book provides a comprehensive overview, making IoT accessible and relevant. The targeted audience will benefit from real-world examples, challenges, and solutions, ensuring a valuable and enriching reading experience.

The journey does not end here. The future of IoT holds exciting possibilities, from integrating Artificial Intelligence and edge computing to the emergence of 6G connectivity.

This comprehensive exploration of IoT is enriched by a collection of ten diverse chapters selected through a rigorous peer-review process. Each contribution offers a unique perspective on the latest developments in IoT, ensuring a well-rounded and insightful journey through this dynamic field. The carefully selected chapters, authored by experts and researchers in the field, provide an in-depth examination of various facets of IoT, from its applications in smart technologies to challenges, solutions, and prospects. This collaborative effort establishes *Internet of Things – New Insights* as a valuable resource, showcasing the richness and depth of contemporary research in the IoT landscape.

Chapter 1, "A Systematic Review on IoT-Based Smart Technologies for Seat Occupancy and Reservation Needs in Smart Libraries at Institution of Higher Learning", examines IoT's transformative impact on academic libraries. Through a systematic review, it navigates the challenges and solutions surrounding Smart Library Seat Occupancy and Reservation Systems, offering clarity on the evolving landscape.

Chapter 2, "Methods for Detection and Prevention of Vulnerabilities in the IoT (Internet of Things) Systems", explores cybersecurity methods for detecting and preventing vulnerabilities in IoT systems. It outlines signature-based and anomaly-based models, providing insights into safeguarding the interconnected world of IoT.

Chapter 3, "Applications of AI and IoT for Advancing Date Palm Cultivation in Saudi Arabia", highlights the synergy of AI and IoT in advancing agriculture, specifically date palm cultivation in Saudi Arabia. The chapter delves into applications in precision irrigation, smart systems, cold storage management, and quality optimization, offering a glimpse into the future of smart agriculture.

Chapter 4, "Application of Internet of Things (IoT) in Biomedicine: Challenges and Future Directions", bridges the gap between IoT and health care by addressing the slow adoption of IoT in health care, exploring the Internet of Medical Things (IoMT) and its applications in biomedicine.

Chapter 5, "A Survey on IoT Fog Resource Monetization and Deployment Models", explores the economic aspects of IoT, focusing on fog resource monetization. Delving into hybrid architectures and multi-hop IoT strategies offers insights into optimizing IoT deployments, particularly in remote environments.

Chapter 6, "Design and Implementation Smart Petrol Station Using Internet of Things", highlights the automation of petrol stations. This chapter pioneers the design of a Smart Petrol Station, leveraging RFID technology, Wi-Fi modules, and advanced encryption protocols. It exemplifies the integration of IoT for operational efficiency and security.

Chapter 7, "Hybrid Architectures to Improve Coverage in Remote Areas and Incorporate Long-Range LPWAN Multi-Hop IoT Strategies", explores hybrid architectures for improving IoT coverage in remote areas. By combining various technologies and emphasizing Low Power Wide Area Networks (LPWANs), this chapter charts a course for enhanced IoT connectivity in challenging environments.

Chapter 8, "IoT and Energy", examines how IoT revolutionizes the energy sector. The chapter explores real-time data collection, AI integration, and edge computing, offering a roadmap for a sustainable and technologically advanced future in energy management.

Chapter 9, "Effective Screening and Face Mask Detection for COVID Spread Mitigation Using Deep Learning and Edge Devices", addresses the urgent challenges posed by the COVID-19 pandemic. The chapter illustrates innovative solutions for effective screening and face mask detection through deep learning and edge computing, contributing to global efforts to mitigate the virus's spread.

Finally, Chapter 10, "Smart Healthcare at Home in the Era of IoMT", presents the integration of IoT with the IoMT in smart homes utilizing Digital Twins, machine learning algorithms, and human action recognition to enhance health delivery. The chapter introduces Smart Healthcare at Home (SHAH), offering a personalized and convenient healthcare experience. This chapter provides valuable insights into the technical components, design rationales, and challenges of implementing SHAH systems.

New Insights into IoT: As we conclude this exploration, we must recognize IoT's new insights into the technological landscape. The potential of IoT lies not only in connecting devices but also in the vast opportunities for data collection, analysis, and intelligence. The convergence of IoT with other cutting-edge technologies opens new horizons for innovation, offering a glimpse into a future where interconnected devices seamlessly communicate, collect meaningful data, and contribute to the growth of intelligent ecosystems. This book is your guide to understanding the present and envisioning the limitless possibilities that lie ahead in the ever-evolving landscape of the Internet of Things. Embark on this journey and discover the transformative power of IoT.

Maki K. Habib
The American University in Cairo,
New Cairo, Egypt

Chapter 1

A Systematic Review on IoT-Based Smart Technologies for Seat Occupancy and Reservation Needs in Smart Libraries at Institution of Higher Learning

Magauwane Reneilwe Maepa and Michael Nthabiseng Moeti

Abstract

The introduction of industry 4.0 technologies, including artificial intelligence (AI), the Internet of Things (IoT), and other cutting-edge technological developments, has completely transformed traditional library practises in higher education. Despite the topic's unquestionable importance, the main objective of this chapter is to address the effects of IoT technology and it is inconsistent and dispersed. However, there are various challenges, such as accurate reservation methods, real-time seat occupancy tracking, and reservation time estimation. To discover, compare, and characterize current investigations in the Smart Library Seat Occupancy and Reservation system (SLSORS), this proposed book chapter examined articles published between 2016 and 2022. In the SLSORS, we will also give a thorough taxonomy and perform a technical analysis of the articles. This provides the much-needed clarity regarding the problems associated with SLSORS and their available literature-based solutions. The fundamental taxonomy is framed by the reservation security, seat reservation, seat selection, and seat availability criteria. Thus, the benefits and drawbacks of the selected approaches are also offered, along with a full comparison of evaluation methodology, evaluation tools, and evaluation metrics. Furthermore, this chapter incorporates all processes and method for SLSORS and drew attention to the ongoing challenges that this chapter is seeking to address.

Keywords: Internet of Things (IoT), radio frequency identification (RFID), force sensitive resistor (FSR), smart library, seat occupancy and reservation, wireless sensor network (WSN)

1. Introduction

The Internet of Things (IoT) has grown more popular and accepted in the library industry to improve operations, thanks to the increasing use of smart devices and fast networks. As a result, the use of Internet of Things (IoT) in libraries has raised discussion in the academic community. Devices may now connect directly with one another and with the cloud thanks to IoT, which improves the delivery of services in businesses. The requirement for real-time information and high-quality data is gradually growing, particularly for seat occupancy status and reservation in the libraries, as more students are admitted each year. Smart technologies are also being used in every aspect of library services and workflows. To improve the service experience for users, libraries have shown persistence in implementing new technologies.

IoT is the collective term for any connected devices that use the Internet to collect data that can then be used to automatically monitor and operate objects without the need for human interaction. Most studies found that numerous new technology inventions trending nowadays are more advantageous because they can really simplify our lives. Of which one of the foremost drivers of the future smart spaces is the Internet of Things (IoT) [1–3]. Boboc and Cebuc [4] define IoT as the sum of devices interconnected over the Internet, with data collection ability to monitor and control things remotely without human intervention. Therefore, the application of IoT will enhance the value of education process in the education environment because it will allow students to learn quickly [5].

Similarly, Abuarqoub et al. [1] also describe IoT as a system that connects daily things embedded with electronics, software and sensors to the internet enabling them to gather and exchange data. According to Bayani et al. [2], IoT is utilizing smart features of Radio Frequency Identification (RFID) and Wireless Sensor Network (WSN) technologies to change everyday life. Hence, its objective is to enhance our everyday devices and appliances to be less sophisticated, automated, flexible and highly accessible at any time, from anywhere, by any user across the world [1].

According to Brian et al. [6], the principle of developing a connected library system where the user can use their mobile phone to connect to the library system is advantageous and beneficial since almost everyone nowadays has a smartphone device. Upala and Wong [7] and Daniel et al. [3] concurs that IoT can connects numerous devices to expand operational efficiency, real-time visibility, and user learning experiences because of its outstanding potential in the education sector. They revealed that new research opportunities and feasible solution are possible through the development of smart library system utilizing IoT.

It is therefore undeniable that the application of IoT promises a brighter future for libraries [8], but scholarly research is needed to determine how well libraries will embrace this trend considering its prospects and challenges. Numerous research has been done on the implementation of smart-library seat occupancy and reservation to provide students with the luxury of checking seat occupancy status and reserving seats online without librarian interaction Maepa and Moeti [9] utilizing various IoT technologies. And also challenges, drivers, and obstacles were also discussed from a variety of angles Daniel et al. [3] have all conducted several previous literature review-based studies to evaluate the development, trends, gaps, and future research direction in the IoT sector. Although numerous studies have examined IoT-based research from various angles and for various amounts of time [1–3, 7, 9]. Even though the use of IoT in libraries is growing and that there is an increase in research into the topic [10], a study that looks at libraries in developing nations reveals a knowledge gap.

To ascertain the validity of the aforementioned assertions, it may be helpful to conduct a systematic literature review (SLR) study to see if there have been any empirical studies that examined IoT and the technologies that it enabled for seat reservation needs in smart libraries. Therefore, the objective of this book chapter is to conduct an SLR in order to (i) ascertain whether empirical studies have been conducted that address the effects of IoT technology and its inconsistent and dispersed nature for seat occupancy and reservation needs in smart libraries, and (ii) discover, compare, and characterize current investigations in the SLSRS and (iii) determine from which databases such studies, if any, are extracted. Even though there are existing reviews that looked at the emerging smart library systems using IoT technologies, this research is different as is only focusing thoroughly on the element of seat occupancy and reservations to improve library usage as a step to enhance academic excellence and time management. This book chapter seeks to respond to the following research queries (RQs) by reviewing the available research.

- What are the benefits of IoT-Based smart library seat occupancy and reservation systems?
- What are the existing technologies and methodologies that enable IoT-Based seat occupancy and reservation in smart libraries?
- What are the main unresolved problems, potential developments, and difficulties in IoT-based smart library seat occupancy and reservation systems?

This Chapter used primary research publications to conduct an in-depth investigation into the present drawbacks and advantages of smart library seat occupancy and reservation systems and to better comprehend the various models put forth by various writers. Illustrated are the various holes and restrictions in the current IoT-based seat occupancy and reservation systems. Consequently, the chapter excludes other library management systems providing support for teaching and learning activities such as obtaining and reserving library books.

2. Literature review

This section begins with a study of the Internet of Things and its characteristics before highlighting the shortcomings of the present smart library systems using IoT technology. This will assist identify the gap that exists in the systems already in use. Before illustrating the quality of seat occupancy and reservation, a description of SA university library operations, smart systems, and IoT-based smart library systems is given. By highlighting the gap in the literature, outlining the shortcomings of the current seat occupancy and reservation systems, and concluding with a summary of the chapter, related works are accessed to provide a thorough understanding of the IoT-based smart library seat occupation and reservation systems currently in use.

2.1 The nature of the Internet of Things

The Internet of Things (IoT) is the trending technology that simplifies lives nowadays and it also enables the real, digital and virtual convergence to develop smart surroundings that make energy, transport, cities and many other areas more intelligent [4]. Internet of Things describes a system that connects daily things embedded with electronics, software and sensors to the Internet enabling them to gather and exchange data [1]. According to the Jayawardena et al. [11], four communication models, namely device-to-device, device-to-cloud, device-to-gateway and back-end data sharing, affect the IoT. These four communication models describe how IoT objects connect with one another to exchange data. Organizations can effectively leverage the adaptability of IoT connections to suit their various purposes by utilizing the various communication methods. The Internet of Things (IoT) is the connecting of common objects or things for interaction and data sharing for planning, processing, or decision-making, as can be inferred from the aforementioned.

According to Boboc and Cebuc [4], all tangible and intangible things in our daily life are predicted to be connected to the Internet purposely to advance the way people live, work or interact. Nevertheless, Goyal et al. [12] stated that IoT affords all tangible and intangible things with unique identifiers (UIDs) and the ability to transfer data over a network without needing human-to-human or human-to computer interaction. Hence, its objective is to enhance our everyday devices and appliances to be less sophisticated, automated, flexible and highly accessible at any time, from anywhere, to any user across the world [1]. For this reason, IoT has bridged across various application fields, and it combines many technologies [13].

Boboc and Cebuc [4] defined Internet of Things as an innovative technology that incorporates billions of smart objects into our daily life to enhance social, technical, and economic benefits. According to Mohamed et al. [5], the application of Internet of Things will enhance the value of education process in the education environment because it will allow students to learn quickly.

2.2 IoT technologies

IoT links heterogeneous items together, using embedded systems, such as wireless connectivity, computers and wireless smart sensors. According to Jha et al. [14], there are several technologies utilized by IoT such as Internet protocols, communication technologies, CPS, WSN, RFID, and condition awareness [15]. They further categorized devices like storage devices, servers for computing data, security devices for protection, control devices, devices for sensing, capturing, and generating data and some portable devices as the hardware part of IoT. Similarly, Alam et al. [16] highlighted the fact that a few of the technologies related to IoT-enabling include actuator, radio-frequency identification (RFID), wireless sensor network, Near-Field Communication (NFC), M2M communication, and IPv6 over Low-Power Wireless Personal Area Networks (6LoWPAN).

Furthermore, Goyal et al. [12] stated that IoT devices which are utilized for monitoring and tracking comprise of small silicon chips with finest processing capabilities. Communication technologies like GPRS, GSM, Mesh Network, LTE, ZigBee and Wi-Fi are utilized. Contrasting to this, technologies such as NFC, RFID, GNSS & BLE are employed in tracking. However, for detecting vibration, pressure, temperature and humidity, it uses sensors powered by small battery, short burst power bank and solar energy [12].

According to Bansal and Kumar [17], IoT works with emerging technologies like big data, artificial intelligence, Wi-Fi, networking technology and sensing instruments because it is an analytical system. This supports the idea that IoT works best when combined with a variety of technological solutions. However, Albishi et al. [18] noted that the majority of IoT frameworks call on technologies including networking setups, cloud computing setups, internet, and other software applications. Hence, they recognized situation awareness and cognition, autonomy, cloud computing, semantic technologies and semantic technologies as linked future internet technologies for the Internet of Things.

Berek [19] stated that the RFID technology has been used in libraries for years and since its appearance, there has been remarkable solutions applied in the libraries field according to their special requirements. Libraries are gaining new opportunities owing to the continuous technological growth. In accordance with Li [20], RFID is a wireless, non-contact technology that transmits data from a tag connected to a device to track and identify it. Hence, Misra et al. [21] noted that the use of RFID in IoT will support wireless connectivity of small devices and help maintain track of smart objects within the IoT framework in real time. Thus, Radio frequency waves are used for data exchange in the IoT and RFID technology combination to improve device connectivity. Nevertheless, businesses utilizing the IoT framework can connect various kinds of products devoid of transducers like wireless sensors and actuators. As a result, organizations like libraries can set up self-service features, maintain security and keep an eye on operations thanks to RFID.

2.3 Characteristics of IoT

IoT describes a system that connects daily things embedded with electronics, software and sensors to the internet enabling them to gather and exchange data [1]. According to Mohamed et al. [5], IoT is one of the internet phenomena that has successfully disrupted our daily life. Hence, it has grown more popular and accepted in the institutions of higher learning to improve operations, thanks to the increasing use of smart devices and fast networks [3]. Thus, IoT has made it possible for devices to link directly to one another and to the cloud, which improves service delivery and management strategic planning in businesses [12]. Hence, Sundaravadivazhagan et al. [22] and Bayani et al. [2] characterized IoT as follows:

- **Connectivity:** Internet connectivity is attached with in the devices and sensors.
- **Communication:** Everything is unified with comprehensive evidence and communication structure.
- **Things related services**: IoT is capable of traditional and non-traditional computer things related services.
- **Security:** IoT may be transmitting sensitive data, it is very significant to give data privacy and security.
- **Energy Efficient:** The IoT devices should have power backup.
- **Sensing:** It is an important supporting device in IoT.
- **Heterogeneity:** The IoT devices based on hardware and network platforms.
- **Dynamic Environment:** The IoT devices support a dynamic environment.
- **Enormous scale**: The IoT technologies support to control a greater number of devices which interact.

These characteristics are depicted as follows (**Figure 1**).

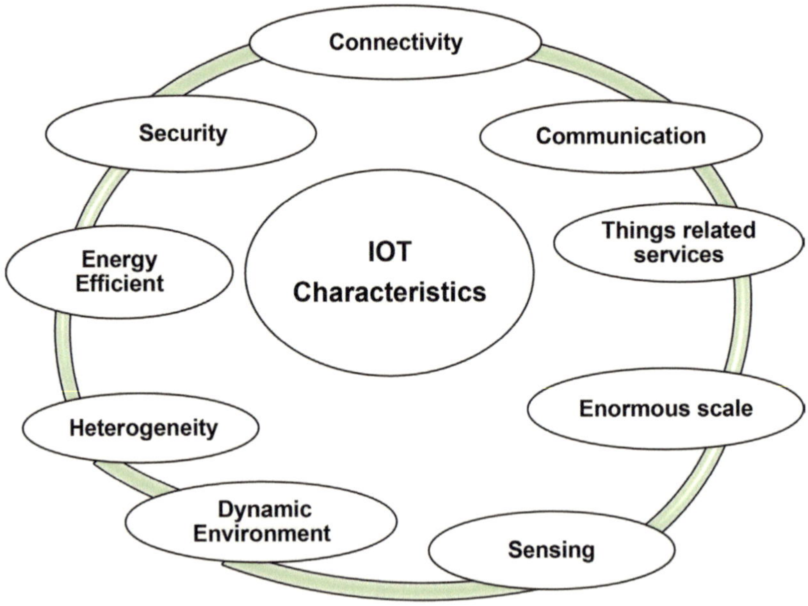

Figure 1.
IoT characteristics (adapted from: [2, 22]).

2.4 IoT applications

The Internet of Things (IoT) applications are categorized into a wide range of application domains, including wearables, smart homes, smart cars, smart infrastructure, smart healthcare, agriculture, manufacturers, supply chains, logistics, social and business applications [23]. As more devices connect to the Internet and become IoT enabled, the number of IoT applications is growing [24]. Today's commonplace applications make use of smart devices. They can 'speak' with one another and exchange vital information and data [10]. Hence, the adoption of traditional distributed system programming methods is hindered by the unique characteristics and needs that each application area possesses [23].

2.5 Smart systems

In the academic community, the term "intelligent" or "smart" systems refers to a wide notion with the aim of maximizing productivity through the application of cutting-edge information, communication and computing technology [1]. The advancements in general IoT research provided the ability to utilize intelligent systems in industrial applications. The idea of "smart" in this context, has more in common with Industry 4.0 than with conventional logic. It includes complex logical operations and algorithms; it is not merely limited to simple logical operators. While describing systems as "smart," there have been some misconceptions in this area of research. Often, researchers will classify a wireless or automated system as "smart". A machine that relies on input signals, range comparisons, triggers and output, cannot be referred to as intelligent and is merely an automated system. The use of smart techniques in emerging systems is assisting in reducing times, the demand for manpower, and the level of skill required to maintain the systems and improve the quality of the output. Using cutting-edge learning methods, such as machine learning, IoT and AI can strengthen and broaden this idea [25].

There are several practical uses for IoT technology, including Industry 4.0, smart agriculture, smart cities, smart transportation, smart homes, eHealth, and wearables [25–27]. According to Boboc and Cebuc [4], all tangible and intangible things in our daily life are predicted to be connected to the Internet purposely to advance the way people live, work, or interact. Hence, Industry 4.0's objective is to enhance our everyday devices and appliances to be less sophisticated, automated, flexible and highly accessible at any time, from anywhere, to any user across the world [1].

2.6 Development of the IoT-based smart library

Naturally, the term "smart library" implies that libraries have developed to the point where they are now a crucial element of smart cities or smart university campuses [19]. Smart technologies are also being used in every aspect of library services and workflows. To improve the service experience for users, libraries have shown persistence in implementing new technologies. As a result, the use of the Internet of Things (IoT) in libraries has raised discussion in the academic community. Devices may now connect directly with one another and with the cloud thanks to IoT, which improves the delivery of services in businesses. It is undeniable that the application of IoT promises a brighter future for libraries [8], but research is needed to determine how well libraries will embrace this trend, considering its prospects and challenges. Even though the use of IoT in libraries is growing and that there is an increase in research into the topic [10], a study that looks at libraries in developing nations reveals a knowledge gap.

2.6.1 Components of an IoT-based smart library

The traits and elements of a smart library can be divided into three categories, namely smart people, smart services and smart technology [28]. These elements working together, is a prerequisite for a smart library. At the same time, the smart library cannot be realized if any one of the following conditions are not met:

- **Smart technology** - The library accepts the technological advancements requested by its patrons because they are already prevalent in all facets of modern life.

- **Smart services** - To meet user needs, services should be planned and implemented. At this level, operations can be user-friendly and service oriented.

- **Smart People** - At this level the library patrons in addition to the personnel are also considered.

Users must demand the possibilities provided by new technologies, but librarians must also be able to explain the new method of working to library patrons and educate them about it, in addition to simply understanding it themselves. These elements interact, but they also reinforce and strengthen one another in all directions. The component parts work together to create a complicated whole.

The availability of technology is one of the elements that defines smart libraries [10]. The way services are organized will be affected by how technology is adapted and used in other fields. The online systems of the library are made possible by a variety of technologies, including wireless systems, RFID, LED and the Internet of Things (IoT) technology [13]. Although it may be argued that these technologies alone do not form a smart library, it is impossible to develop the services that today's library customers have come to expect without them [10].

2.6.1.1 Challenges of IoT in libraries

The current condition of IoT in libraries in developing nations varies from one library type to another and from one country to another, even though technological maturity is a factor in adopting improvements [3]. According to Patel et al. [28], optimal resource management is still a difficulty, given that IoT infrastructure is unevenly spread globally and is focused on high-income nations, even though the integration of intelligent technology offers limitless potential in precise university libraries. This suggests that the existing level of universities involvement in IoT deployment is insufficient to ensure the innovation is adopted as best as possible [10]. Hence, Bayani et al. [2], mentioned that universities should build up systems that encourage the use of new technologies to assist technical advancements. Their assistance in assuring a steady electrical supply, lowering ICT tariffs, and guaranteeing greater network bandwidth with improved connectivity in pub spaces would therefore be expected to influence libraries in poor countries to adopt IoT [28].

According to Liang [10], high security and privacy standards are necessary for IoT, which is why they have turned into elements influencing its acceptance. The author pointed out that while using the IoT, libraries need to warn users about the potential vulnerability of their network, hardware and software. Additionally, there is a risk that some of these devices or connecting tools could have user interface vulnerabilities when multiple devices are interconnected or made to perform interactively [23]. This creates significant privacy and security concerns for IoT applications in libraries. Libraries must therefore implement measures to protect their infrastructure from any risks and hazards posed by IoT applications [10, 23].

The adoption of a certain technology used in IoT may be impacted by physical risk, which is a person's concern that it could be harmful, dangerous, or unhealthy, according to Mani and Chouk [24]. Therefore, when library staff and customers perceive using IoT as being of high risk, it will have a negative impact on its uptake.

2.6.1.2 Overview of the library seat occupancy and reservation systems

According to Daniel et al. [3], the IoT model has been espoused in all study fields with emphasis on the connection of everything to the Internet. The Internet of Things is shifting generations from predictable systems to SMART systems and is mostly employed in urban areas [29]. Nevertheless, it has some implementation difficulties, such as the cost of IoT devices, its development and outline, technical standards and flexibility [7]. However, as declared by Daniel et al. [3], the adaptation of these technologies in the library system eases challenges faced by management concerning to self-servicing, monitoring and tracking of resources in the library.

This overview of the library seat occupancy and reservation systems further analyses the applications and functionalities of the existing IoT-based seat occupation and reservation systems and identifies the possible methods, tools and techniques for implementing an IoT-based smart library system.

Most studies found that the use of numerous new technological inventions devised nowadays, are more advantageous because they can really simplify our lives [30–32]. One of these numerous new technological inventions and foremost drivers of the future smart spaces, is the Internet of Things (IoT) [1–3]. Boboc and Cebuc [4] define IoT as the sum of devices interconnected over the Internet, with data collection ability to monitor and control things remotely without human intervention. Therefore, the application of IoT will enhance the value of the educational process in the education environment because it will allow students to learn quickly [5].

Similarly, Abuarqoub et al. [1] describes IoT as a system that connects daily things embedded with electronics, software and sensors to the Internet, enabling them to gather and exchange data. According to Bayani et al. [2], IoT is utilizing smart features of Radio Frequency Identification (RFID) and Wireless Sensor Network (WSN) technologies to change everyday life. Hence, its objective is to enhance our everyday devices and appliances to be less sophisticated, automated, flexible and highly accessible at any time, from anywhere, by any user across the world [1].

According to Brian et al. [6], the principle of developing a connected library system where users can use their mobile phones to connect to the library system, is advantageous and beneficial, since almost everyone nowadays has a smartphone device. Upala et al. [7] and Daniel et al. [3] concur that IoT can connect numerous devices to expand operational efficiency, real-time visibility and user learning experiences, because of its outstanding potential in the educational sector. They reveal that new research opportunities and feasible solutions are possible through the development of a smart library system utilizing IoT.

2.6.1.3 Using the RFID technology system in libraries

There are many places that are currently equipped with sensors for temperature detection, traffic automation, and vehicle and UAV autopilot [2]. Thus, IoT and its supporting technologies, such as M2M communications, V2V communications, RFID and NFC may power a variety of applications. Nevertheless, Bayani et al. [2], stated that IoT has a promising future with the increasing connectivity of devices and objects.

RFID is a barcode replacement that employs tiny microchips in tags to store and send extensive data in the database about the item being tagged [33]. Nevertheless, Hirekhan [34] described it as the usage of microchips and library cards, enabling clients to check out items by going through a self-service station equipped with an antenna that transmits low frequency radio waves. Hence, it is pointed out that it can be used to identify, track, sort, or detect library holdings at the circulation desk and during daily stock maintenance [34]. The notion of RFID can be compared to that of an electronic barcode. An RFID system primarily consists of the following four parts:

- RFID transponders or tags that have been electronically programmed with special data.
- Tag-querying sensors.
- An antenna.
- Server on which the integrated library software's interface program is loaded.

This technology made up of smart RFID labels, related hardware and software, gives libraries a better way to manage their collections while also giving their customers better service [35]. RFID-tag reading does not require manual interactions. The functions and advantages provided to the library with the utmost care using RFID technology include the reducing of manual involvement, reduced manual errors, and enabling quick book issuing, reissuing and searching [36]. Below are the advantages of RFID system:

Rapid Charging/Discharging: The usage of RFID shortens the time necessary to complete circulation processes. The information from RFID tags can be read much more quickly than from barcodes, and multiple objects in a stack can be scanned at once. This accounts for the biggest time saving. Despite being originally unreliable, the anti-collision algorithm that enables the charging or discharging of a whole stack currently seems to be performing smoothly since then.

Patron Self-Charging/Discharging Simplified: For patrons' self-charging, there has been a noticeable improvement because they are not needed to meticulously position materials within a specific template and they are able to charge multiple products at once. When customers self-dispense, staff no longer have to do the work. The installation of backdrop readers significantly relieves the staff's work.

High Reliability: You may rely on the readers. To detect the items leaving a library, certain RFID systems incorporate an interface between the exits sensor and the circulation system. If a customer hurries out of the library without being too closely observed, the library would at least be aware of what had been taken. The library will be able to identify who takes an item out without paying for it if the user card also includes an RFID tag. This is accomplished by creating a bit that serves as the "theft" bit and turning it off, both while charging and discharging.

Long Tag Life: Finally, RFID tags have a longer lifespan than barcodes since nothing touches them. Most RFID providers state that a tag can withstand at least 100,000 transactions before perhaps needing to be changed [37].

2.7 Gaps in the literature

Research on finding a solution to the issue of seat occupancy has increased recently. Upala et al. [7] constructed an IoT setting in an academic library, embedded with a security parameter of "face recognition" for user identification within library management. This was done in order to support library space management; such as study room occupancy service by using IoT applications. Their proposed solution was considered a quality intelligence system as it afforded participants such as students, staff and authorized users with a secured real-time view of library assets, to justify study room or conference room usage. However, their focus was on seat occupancy which does not include seat reservation. They define libraries as vital fragments of the educational system, used to advance our knowledge. Hence, several societal backgrounds have advanced because of the growth of IoT development of which the traditional library system is one [2].

Similar to Hoang et al. [31], Yahaya et al. [32] constructed a seat occupancy detector system, with both systems in the same context; using capacitance sensors as an innovative method to differentiate the seats occupancy by either subjects or objects. Both their systems were developed using a Raspberry Pi as main controller, integrated to a Wi-Fi module, along with a capacitance sensor chip. Their projects were intended to ease the congestion, especially during or before the test week and exam time, as these periods are a recurring problem for most libraries. Although, their results were unreliable as they failed to distinguish whether a seat has been occupied by a person or an object, they nevertheless showed the seat occupancy status. Hence, they pointed out that there is a need to advance their analytical algorithms to accurately distinguish the occupancy detection states.

Torres and Paul [31] employed a library seat occupancy counter to gather daily counts over time that would specify patterns in student preferences for study spaces and to also use that information to advance several library services. The researchers employed Microsoft Paint (MS Paint), Excel and Qualtrics (a survey and researching tool) [31]. However, their study does not report on how students engaged with space, and it also required student workers to manually enter the number of students occupied seats and their preference to get daily counts. Their methodology is not reliable as it could lead to students entering erroneous data which could produce invalid counts. Hence, they suggested that there is a need to develop a real-time seating visualization that would be available via an app, which could enable library users to check and reserve available seat based on their preference [31].

The issue of managing library seats was also addressed by Daniel et al. [3]. They described the planning and implementation of a solution that combined hardware and a web application to allow students and librarians to verify the identification of library seat occupants and the status of library seat occupancy from any location at any time over the Internet, using their devices or the display system at the library door. In the library, the prototype system of Daniel et al. [3] significantly reduced the amount of time students spent looking for a seat and making or receiving phone calls to other students. Nevertheless, their Web Application did not have any integrated privacy or visibility features.

Nevertheless, library users could not reserve a seat; their system only permitted the luxury of seat occupancy and monitoring. Daniel et al. [3] developed a prototype of a smart library seat occupant and occupancy information system, made of pressure (force-sensing resistor) and RFID sensors for library seats, which send real-time seat utilization status to the web application. The researchers recommended including a feature that would permit seat occupants to take a little break (reserve) while in the library in the future's smart library systems [3]. This is the gap that the researcher is hoping to fill with the current research.

Several studies were conducted in other countries regarding this most relevant concept of 'smart library' [31–32]. Researchers such as Brian et al. [6] and Bayani et al. [2] focused on helping students to find available books and reserving them online. However, most other studies focused on seat occupancy and seat monitoring [3, 31–32]. Nevertheless, to this date, as far as could be established, no one has attempted to investigate the smart library seat occupancy and reservation in SA.

According to Daniel et al. [3], "there is a need to integrate the University map into the web application to enable students within the university premises to reserve seats with their Unique Identification (UID), based on the estimated time of arrival at the library". This means that there is a gap in the literature and research on seat reservation based on estimated arrival time and reliable seat occupancy status, which needs to be filled.

Based on the main problem discussed in the problem statement, there is a need of inventing better, efficient, and convenient ways of saving time and energy to enhance academic services and excellence. Since technology is growing continuously, we should embrace it and advance with it too. As a result, several contemporary technologies have been developed to integrate library seats with smart devices that enable library users to locate the study sections with open seats, thus resolving the issue of seat availability or reservation. Most commonly, these technologies make use of Internet of Things (IoT) and wireless sensor networks. The below **Table 1** shows a summary of different research methods, their results, research gaps and similarities on smart library seat occupancy and reservation systems as explained above under the heading named gaps in the literature. **Table 1** content described as follows:

Reference source - displays the names of the author(s) involved in the research study

Research problem – shows the research problem that the author(s) were trying to solve.

Research methods – a list of all the research methods the authors used to address the specific research problem.

Research result – demonstrates the solution the authors came up with for the issue they identified.

Reference Source	Research Problem	Research Method(ology)	Research Result	Research gap(s) and similarity/difference
Upala and Wong [7]	To construct an IoT setting in academic library embedded with security parameter of "face recognition" for user identification within library management; to support library space management such as study room occupancy service using IoT applications.	LBPH face recognition algorithm ThingSpeak channel Up-squared board ultrasonic sensor ThingSpeak cloud. SQLiteStudio database Python	The resulting logs have all the specific details about individual sensor data entry. The effectiveness of the occupancy sensor detection. No privacy and security on user information.	Their system only permit the luxury of seat occupancy and monitoring. Qualitative and quantitative data can be collected to learn what students like to do in the library space and provide deep understanding for library space management.
Daniel et al., [3]	the design and execution of a resolution merging hardware and web application that permitted students and librarians to authenticate the identity of library seat inhabitants and occupancy status of the library seats anywhere anytime over the Internet using their devices or display system at the library entrance.	Prototype using FSR (force-sensing resistor) sensor and Radio Frequency Identification (RFID) reader Wi-Fi module (node MCU) – IoT gateway Database PHP API	Their system only permit the luxury of seat occupancy and monitoring. Usability: 93% Performance 99% No privacy	Their system only permit the luxury of seat occupancy and monitoring. Library users could not reserve a seat
Hoang et al. [31], Yahaya et al. [32]	To construct a Seat Occupancy Detector system both in the same context; using capacitance sensors as an innovative method to differentiate the seats occupancy by either subject or object.	Raspberry Pi as main controller, integrated to Wi-Fi module, along with a capacitance sensor chip	showed the seat occupancy status.	their system only permit the luxury of seat occupancy and monitoring To distinguish on whether the seat has been occupied by a person or an object

Reference Source	Research Problem	Research Method(ology)	Research Result	Research gap(s) and similarity/difference
Torres and Paul [30]	employed a library seat occupancy counter to gather daily counts over time that would specify patterns in student preferences for study spaces and to also use that information to advance several library services.	Microsoft Paint (MS Paint), Excel, and Qualtrics (Survey and researching tool)	student workers manually enters the number of students occupied seats and their preference to get daily counts	their system only permit the luxury of seat occupancy and monitoring To develop a smart library system which will enable library users to check and reserve available seat based on their preference

Table 1.
Summary of different research methods, their results, research gaps and similarities.

Research gap(s) and similarity/difference – displays the remaining gaps surrounding the topic, what still needs to be done, how their research differs from and is comparable to the current research.

3. Review planning and research methodology

3.1 The systematic literature review (SLR)

A systematic literature review (SLR) was conducted to find out what other researchers found about the smart library seat occupancy and reservation systems. A SLR concentrates on an analysis of several investigations within a particular field of research [38]. It is also a method for locating, analyzing, and synthesizing all data pertinent to a specific research question, topic, or phenomenon of interest [39]. Additionally, SLR is an approach that streamlines the steps involved in compiling, categorizing, and rating literature in a review area [40, 41]. Based on the goal of this investigation, which is to identify relevant results in current research and make recommendations for future research, a systematic review was deemed appropriate in this study [42]. Systematic reviews are becoming more and more prevalent across all disciplines and in sectors that combine IT and Academia [43]. Professionals and academics utilize systematic reviews to keep current in their disciplines, as a foundation for creating technology guidelines that can be applied to other sectors such as libraries [44]. It is therefore utilized as a tool for finding, assessing and interpreting research pertinent to a certain research issue or an area of interest. By highlighting gaps and outlining potential research subjects, a systematic review can considerably advance our understanding of a certain field of study [45]. Researchers that use this strategy must follow the principles and recommendations of systematic reviews [39]. The systematic review will be highly efficient if it is started utilizing a strategy to find, pick, and evaluate the pertinent literature [46]. According to Boell and Cecez-Kecmanovic [47], the systematic procedure should be rigorous, clear, objective, and repeatable.

The structured methodology Watson [48] suggested, which explicitly lays out the procedures and techniques for literature searching, provides the foundation for this systematic review. The process of a systematic review protocol consists of the following phases: planning, execution, and reporting. During the planning stage, rules and criteria are implemented, such as determining the necessity of a systematic review, creating a categorization framework, formulating research questions, and formulating research methodologies. The execution process covers the methods of identifying a trustworthy data source for the research questions; conducting a search strategy; conducting the study process; choosing pertinent primary studies (using inclusion and exclusion criteria); evaluating the quality of the studies; and extracting the data. This research's reporting phase comprised categorization of the chosen articles and discussion of the findings. **Figure 2** describes the procedures, regulations, and standards followed for this systematic review.

In research a systematic review finds, groups, synthesizes a comparative approach of enquiry, and promotes information transfer [38]. To achieve this, the researcher reviewed existing research papers, articles, book chapters and journals related to this topic. The questions for the research determining the emphasis and the paper's objectives are listed. Then, an SLR approach is used, which entails planning, carrying out, and recording the issue at hand. This section provides a summary of the planning and conducting phases that were used to carry out this. Finally, the study clearly explains the procedure for reviewing the data and protocol.

3.2 Planning phase

3.2.1 Determining the necessity of a systematic review

Identifying the needs for the systematic review is the first step in the planning phase. The demand for researchers to thoroughly and objectively summarize the information already available about a phenomenon gives rise to the need for a systematic review. Even though there is ongoing study on the roles, advantages, and difficulties of IoT-based smart libraries. To our knowledge, no systematic review has, however, summarized these research results and offered a thorough overview of the research and practice on this subject.

Figure 2.
Systematic review phases.

3.2.2 Creating a categorization framework

The creation of the research review procedure, which serves as a foundation for understanding the current theoretical and practical perspectives on the subject, is the second step of the planning stage. The review protocol for this study details the procedures followed to conduct a particular systematic review. To prevent researcher bias, a predetermined protocol is required. Without a procedure, for instance, the choice of certain studies or the analysis may be influenced by researcher expectations. IoT is used in many different professions. Henceforth, it was required to review only the research data that were currently available and relevant to the issue covered by the study.

The following procedure was followed:

The study-related basic search phrases were compiled into a list. These key phrases were derived from the research topic of the study. IoT-Based seat occupancy and reservation, seat reservation, seat occupancy, and seat monitoring in the context of smart libraries were the keywords.

This process incorporated the identification and comparison of different smart library seat occupancy, monitoring, and reservation systems and their weaknesses. It helped the researcher to discover about the current operations on seat occupancy and reservation systems in SA libraries and the gaps that still need to be filled. To discover, compare, and characterize current investigations in the SLSRS, this study examined 14 articles published between 2016 and 2022.

3.2.3 Formulating research questions

A systematic review's third planning phase, which is also regarded as being one of the most important, is defining the research questions [41]. When it is able to respond to the research questions, a systematic review succeeds in its objectives [42]. The following research questions were put out for this systematic review study:

- What are the benefits of IoT-Based smart library seat occupancy and reservation systems?

- What are the existing technologies and methodologies that enable IoT-Based seat occupancy and reservation in smart libraries?

- What are the main unresolved problems, potential developments, and difficulties in IoT-based smart library seat occupancy and reservation systems?

3.2.4 Formulating research methodologies

The fourth phase of planning is the definition of article selection techniques. The goal of article selection strategies is to locate the primary studies that directly address the research issue. Although they may be improved during the search process, techniques for article selection should be determined upon during the protocol definition to lessen the probability of bias [40]. An integrated search approach was used in this step to cover both a thorough automated search of several internet databases and a manual examination of the chosen articles.

The online databases considered for this systematic review are ACM, Emerald, IEEE, Science Direct, Springer, Taylor and Francis. Additionally, appropriate filtering procedures were applied for each selected database to limit the research findings and reduce duplication [49]. The broad manual review technique was used for the manual review, which comprised examining each research article's title and abstract first [50] before reading the complete text of the papers that were chosen to be excluded.

The backward snowball method was utilized to find items that were missed by the earlier techniques in addition to the extensive automated search and manual assessment. To find new publications, this method used a reference rundown [51]. The first step in the backward snowballing technique was to check the reference list and eliminate any papers that did not meet the important research requirements, including language, peer-review status, publication year, and type of publication. The remaining papers were then included in the study after the duplicate articles had been eliminated.

To be included in this research, research studies, research articles and papers were included or excluded in terms of the criteria listed as follows in **Figure 3**. The chosen studies had to meet all inclusion criteria, with none of the exclusion criteria having been met. These criteria make ensuring that the literature taken into account in the systematic review is pertinent to the study, which helps to provide conclusions that are more precise, impartial, and meaningful. These criteria lower the possibility of bias and errors if they are established and applied correctly. Applying eligibility criteria consistently prevents irrelevant studies from being included in reviews, which can result in conflicting findings.

3.3 Execution phase

The planning phase's tactics were applied during the execution phase to choose pertinent articles for the study. The process of a systematic review execution protocol consists of the following steps: identifying a trustworthy data source for the research questions; conducting a search strategy; conducting the study process; choosing

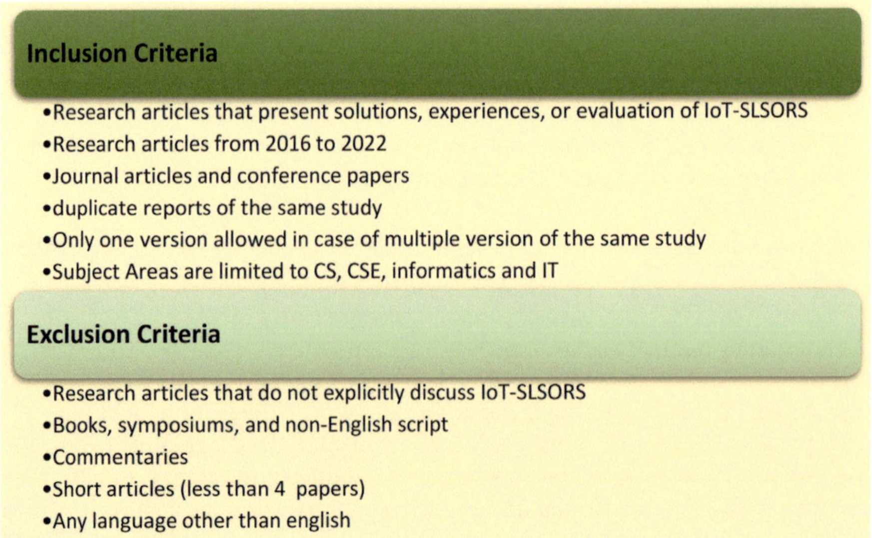

Figure 3.
Inclusion and exclusion criteria.

pertinent primary studies (using inclusion and exclusion criteria); evaluating the quality of the studies; and extracting the data, respectively entailing the following. The rigor of adhering to a pre-established procedure and specific search technique, according to Boell and Cecez-Kecmanovic [47], makes systematic review an effective method. Although Watson admits that effectiveness is crucial in research, he also contends that efficiency is vital. He claims that effectiveness is attained through "synthesizing the literature and revealing the depth of knowledge on an area's critical key concepts and the relationships between these concepts" ([48], p. 185). The main procedures used in the study are described below:

The next sections outline how each step of SLR was carried out to fulfill the set objectives of this study. **Figure 4** Provides a summary of the application of the SLR steps.

Step 1: identifying data sources

Using Google Scholar as the primary search engine and relying on well-known academic publishers like ACM, Emerald, IEEE, Science Direct, Springer, Taylor and Francis (listed in **Figure 5**), the researcher investigated the benefits, limitations, potential developments and challenges using the search string of "IoT-Based seat occupancy and reservation, seat reservation, and seat monitoring in the context of smart libraries."

Step 2: Search strategy

Utilizing specific search phrases from well-known literature in the field of study is the first step in the continual process of identifying search terms [41]. When all of the well-known articles have been located using the aforementioned guidelines, the process is complete. The study's chosen databases include sophisticated search capabilities that let users combine pertinent search terms. The study-related basic

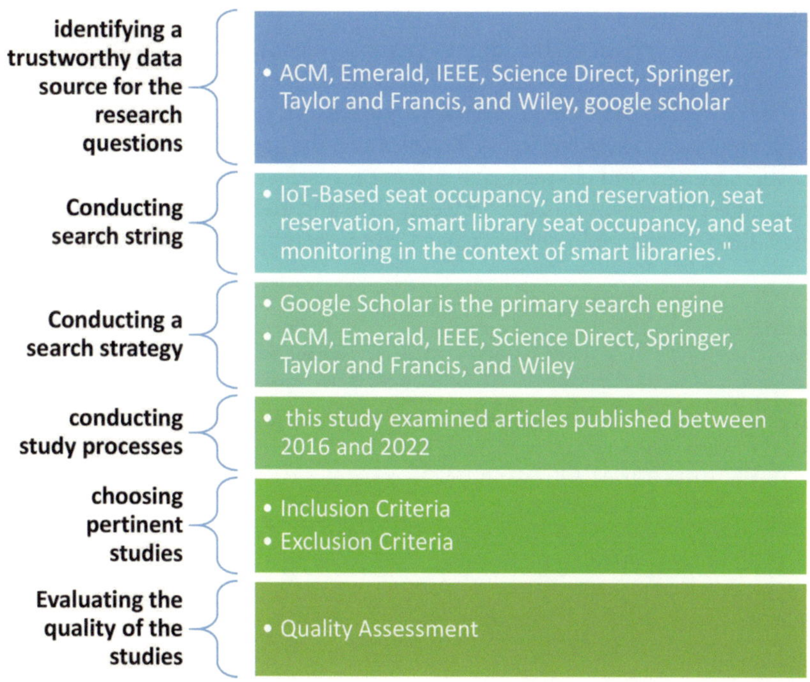

Figure 4.
Summary of the application of the SLR steps.

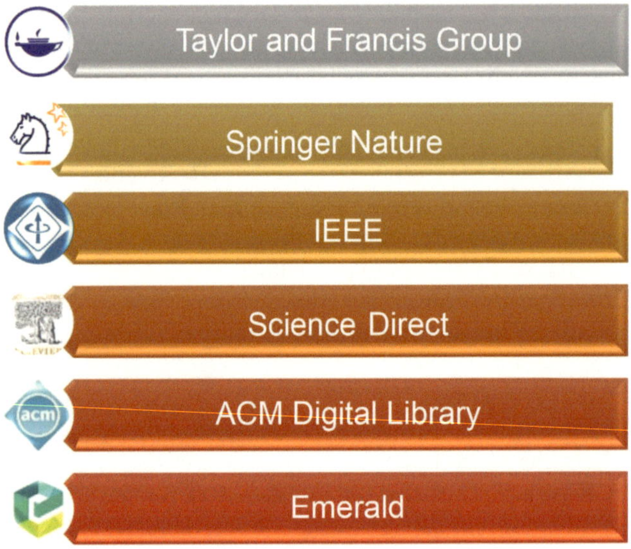

Figure 5.
Databases included.

search phrases were compiled into a list. The following search terms were discovered in this research study: "IoT-Based seat occupancy" OR "IoT-Based seat reservation' OR "seat reservation "OR "smart library seat occupancy "OR "seat monitoring" OR "Seat occupancy" AND "benefits" OR "advantages" AND "methodologies" AND "functionality" (**Table 2**).

Step 3: Study process

This process incorporated the identification and comparison of different smart library seat occupancy, monitoring and reservation systems and their weaknesses. It helped the researcher to discover the nature of the current operations on seat occupancy and reservation systems in SA libraries and the gaps that still need to be filled. To discover, compare and characterize current investigations in the IoT-SLSRS, this study examined articles published between 2016 and 2022. To improve the research outcomes, filtering technologies were used when searching the web databases [44].

Search strings	Online databases searched	Results	Total after Quality Assessment
"IoT-Based seat occupancy" OR "IoT-Based seat reservation' OR "seat reservation "OR "smart library seat occupancy "OR "seat monitoring" OR "Seat occupancy"	Taylor & Francis	50	0
	IEEE	305	6
	Springer	463	0
	Science Direct	117	0
	ACM	41	5
	emerald	7	3
Final results		**983**	**14**

Table 2.
Review search results with search strings.

Several filters were used in this study, including those for the research area (CS, CSI informatics, and IT), year of publication (2016 to 2022), document type (journal articles, conference papers), and language (English).

Step 4: Choosing pertinent primary studies (using inclusion and exclusion criteria)

To ensure that only pertinent papers and other printed sources were used in this study, sources were selected by means of the following broad guidelines, whilst inclusion and exclusion criteria were also identified:

- Only online databases' retrievals of English-language research papers in the form of journals, workshops, conference proceedings and book chapters were considered. The reason being that English is a standard communication language across the world.

- Journal articles were restricted to those that were published between 2016 and 2022. This was done to guarantee that the investigation would utilize current, accessible data.

- Papers that either lacked a clear connection to IoT-Base smart library seat occupancy and reservation, had no connection to the research question, or whose full texts were unavailable, were excluded.

- Duplicate reports of the same study were removed. In cases of multiple versions of an article, only the entire version was added, leaving the others out.

- Articles had to fall into the subject areas of Informatics, Information Technology (IT) as well as Computer Science (CS). This criterion was used to guarantee that the results would be based only on publications from relevant subject areas that acknowledge IoT and its applications in the smart library context. Only those that afford library users with the advantage to view seat occupancy status and reserve seats at their places of comfort.

- The citation information, along with the abstract and keywords, were chosen as part of the dataset extraction process.

Step 5: Evaluating the quality of the studies

The quality score was utilized in this study to examine whether there was a relationship between the primary study's findings and the study's quality. The study also looked at whether certain particular quality parameters, such as sample size and validation method, were linked to the main finding of the study. To reduce bias and increase the validity of the systematic review, it is crucial to evaluate the quality of the primary relevant studies after choosing them. As a result, quality standards were applied to the 14 remaining items. To make sure that study concepts and methodologies were honored, the chosen studies were evaluated in terms of scientific diligence, reliability, accuracy, and appropriateness. The conclusions were examined to determine whether they were focused, unique, relevant, and helpful for upcoming scholars, professionals, and businesses. To provide important and worthwhile contributions to the scholarly community, these requirements were crucial. The chosen studies were grouped based on their primary study objectives, approaches, contributions, and outcomes. This category aids in locating, extracting, classifying, and synthesizing data in response to research questions. The current review chapter was conducted from March 3rd, 2023, to July 30th, 2023, in accordance with the planning phase's specified research procedure. In the initial search using the specified keywords, 983 articles were found. The final 14 research articles satisfied the quality evaluation criteria after completing all the filtering methods described in this stage.

The ultimate number of papers chosen for the current review study is shown in **Table 3**. Particularly, 983 distinct articles were found based on the initial search procedure (keywords). Filters were used to restrict the number of articles to 576. The next step was a manual review by the researchers to find any papers not pertinent to the study. The researchers concentrated on papers that were closely connected to the subject of this research, both conceptual and empirical, during this approach. 510 articles were thus dropped, while 66 were kept as a result. The researchers then read the entire paper, focusing on particular criteria such the aims, the research questions, the description of the data that was gathered, the methodology employed, the approach used to analyze the data, and the presentation of the findings. After reading the complete articles, another 54 were eliminated because they were deemed unimportant, leaving only 12. Then, using the backward snowball method, 8 more articles were added, bringing the total to 20 articles. Finally, 6 papers were eliminated following a review based on the quality evaluation criteria, bringing the total number of publications for analysis down to 14.

3.4 Reporting phase

3.4.1 Categorization of the chosen articles

3.4.1.1 Article distribution by year of publication

The earliest reports on the use of IoT in the libraries date back to 2016 (see **Figure 6**). The year 2019 saw the publication of the most articles (6), while the year 2016, 2017, and 2020 have the least publications (only 1 each year). The fact that

	Automated search method		**Manual Search Method**		**Backward snowball**	**Final Results**
Database	**Keyword Results**	**Apply filter**	**Title and Abstract screening**	**Article Reading**	**Backward snowball Technique**	**Quality Assessment**
IEEE	305	41	34	6	7	6
Springer	463	444	12	0	0	0
ACM	41	38	12	5	8	5
Emerald	7	3	2	1	5	3
Science Direct	117	34	4	0	0	0
Taylor and Francis	50	16	2	0	0	0
Total	**983**	**576**	**66**	**12**	**20**	**14**

Table 3.
Review search results.

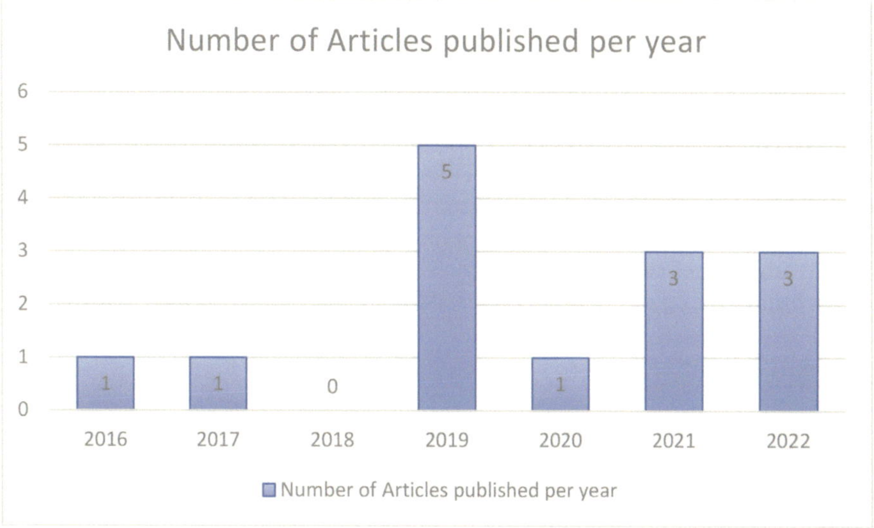

Figure 6.
Number of articles published per year.

most of the articles were released between 2021 and 2022 indicates that this research field has recently attracted attention.

3.4.1.2 Database distribution of articles

The distribution of the chosen articles by database source is shown in **Figure 7**. There were 6 publications found in the IEEE database, 5 in the ACM Digital database, 0 in the Taylor and Francis database, 3 in the Emerald database, and 0 in the Science Direct database.

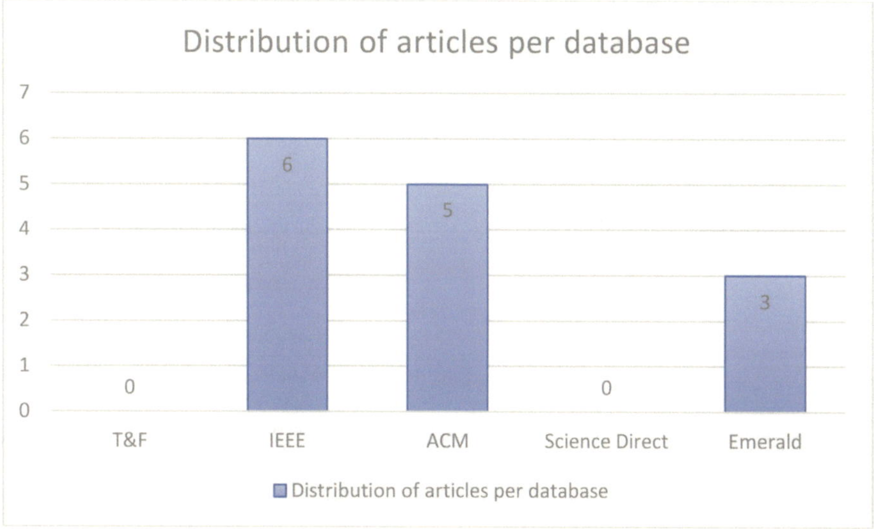

Figure 7.
Distribution of articles per database.

4. Research discussion

4.1 Element 1: benefits

The advantages that libraries can obtain from employing seat occupancy and reservation systems are tied to this element. Time and energy saving, real-time occupancy status, real time seat reservation, increased flexibility, simple and easy to use are some of the advantages it offers to library users [3, 9–53]. Additionally, it provides library staff with advantages such as strategic decision making, develop efficient and effective environment, extracting reports and reducing manual effort (see **Table 4**) [3, 9].

4.2 Element 2: functionalities

Functionalities, in the context of this study, refer to the features that libraries receive through the adoption and use of IoT-based seat occupancy and reservation

Element	Category	Type	Sources
Benefits	Library user	• Saves time and energy • Real-time occupancy status • Real time seat reservation • Increased flexibility • Helps in decision making • Highly portable • Easy to deploy • Effective • Simple • Easy to use • Temporary leave and return after reservation • Improve seat utilization	Maepa and Moeti [9]; Daniel et al. [3]; Zhou [53]; Wang and Wei [52]; Xu et al. [54]; Liu et al. [55]; Lipat et al. [56]; Yang and Wang [57]; Jing et al. [58];Upala and Wong [7]; García-Granja et al. [59]; Hoang et al. [31], Yahaya et al. [32]
	Library staff	• Strategic decision making • Develop efficient and effective environment • Extract reports • Reduce manual effort	Maepa and Moeti [9]; Daniel et al. [3]; Zhou [53]; Wang and Wei [52]; Xu et al. [54]; Liu et al. [55]; Upala and Wong [7]; Hoang Huy et al. [31], Yahaya et al. [32]
Functionalities	Library user	• Check seat availability • Reserve seat	Maepa and Moeti [9]; Daniel et al. [3]; Zhou [53]; Wang and Wei [52]; Xu et al. [54]; Liu et al. [55]; Lipat et al. [56];Yang and Wang [57]; Jing et al. [58];Upala and Wong [7]; García-Granja et al. [59]; Hoang et al. [31], Yahaya et al. [32]
	Library staff	• Manage library seats • View reports • Manage account	Maepa and Moeti [9]; Daniel et al. [3]; Zhou [53]; Wang and Wei [52]; Xu et al. [54]; Liu et al. [55]; Lipat et al. [56];Yang and Wang [57]; Jing et al. [58];Upala and Wong [7]; García-Granja et al. [59]; Hoang et al. [31], Yahaya et al. [32]

Table 4.
Library seat occupancy and reservation systems benefits and functionalities.

systems. The systems allow library users to check the status of the library anywhere and at any time, utilizing their mobile phones or any device with Internet access reducing manual efforts. They ensure that reserved seats in the library study area are secured and that users can reserve seat according to their preference. Everyone is allowed to view the library status and check the availability of seats, but only registered students are allowed to reserve their library seats [3, 9, 53]. These systems act as a medium between the library users and the library staff (librarian). The library staff would have the ability register new library users, view seat occupancy status and produce detailed reports (see **Table 4**).

4.3 Element 3: methodologies

The IoT-based seat occupation and reservation systems are discussed in detail in this Chapter, illustrating various IoT-based seat occupation and reservation systems using different IoT technologies. These IoT methodologies include LBPH face recognition algorithm, Thing Speak channel, Up-squared board, ultrasonic sensor, Thing Speak cloud, SQLite Studio database, Python, RFID, FSR, Raspberry-Pi, Wi-Fi Module, Node MCU, PHP API and capacitance sensor chip. To establish the seat occupancy and rank the options, some approaches used a single procedure, such as the capacitance sensor technology. A combination of methods, including RFID and FSR, were used by other authors to identify the seat occupancy. To find out how many seats were occupied, researchers used a face recognition system, an ultrasonic sensor, and other tools. The gap in the literature is discussed and the shortcomings of these IoT-based reservation and seat occupation systems are highlighted. The adoption of mixed methods enhances performance. Other studies employed RFID and FSR. While FSR detects seat occupancy and transmits the information to the network layer, RFID is responsible for identifying authorized users who made a specific seat reservation (**Table 5**).

5. Delimitation

The focus of the study was based on an smart library seat occupancy and reservation system, which affords students and staff with the advantage to check library seat availability status and reserve seats online. Consequently, the study excludes other library management systems providing support for teaching and learning activities such as obtaining and reserving library books. To ensure that only pertinent papers and other printed sources were used in this study online databases' retrievals of English-language research papers in the form of journals and conference proceedings were considered. Journal articles were restricted to those that were published between 2016 and 2022 and papers that either lacked a clear connection to IoT-Base smart library seat occupancy and reservation, had no connection to the research question, or whose full texts were unavailable, were excluded. Duplicate reports of the same study were removed. In cases of multiple versions of an article, only the entire version was added, leaving the others out. Articles had to fall into the subject areas of Informatics, Information Technology (IT) as well as Computer Science (CS). Only those that afford library users with the advantage to view seat occupancy status and reserve seats at their places of comfort. The citation information, along with the abstract and keywords, were chosen as part of the dataset extraction process.

Source	Methodology	Suggestions/Future research
Maepa and Moeti [9]	DSR IoT RFID and FSR Rasbery pi	enforce security on reserved seats by adopting security access door strategy using RFID and sensor magnetic technologies to attach seats and tables
Daniel et al. [3]	FSR (force-sensing resistor) sensor and Radio Frequency Identification (RFID) reader Wi-Fi module (node MCU) – IoT gateway Database PHP API	Afford users to reserve seat based on estimated arrival time
Zhou [53]; Wang and Wei [52]; Xu et al. [54]; Liu et al. [55]	We chat	use more new intelligent management system technology, improve the overall service and management ability, create a new spatial planning and layout experience, improve the level of Intelligent Library services, more should be the direction of this subject to continue to study
Lipat et al. [56]	Faster R-CNN, RetinaNet, and SSD	
Yang and Wang [57]	cameras TensorFlow.js crawler program, 3D scene	
Jing et al. [58]	Arduino single chip microcomputer	
Upala and Wong, [7]	LBPH face recognition algorithm ThingSpeak channel Up-squared board ultrasonic sensor ThingSpeak cloud. SQLiteStudio database Python	Qualitative and quantitative data can be collected to learn what students like to do in the library space and provide deep understanding for library space management.
García-Granja et al. [59]	building information modeling (BIM)	
Hoang et al. [31]; Yahaya et al. [32]	Raspberry Pi as main controller, integrated to Wi-Fi module, along with a capacitance sensor chip	To distinguish on whether the seat has been occupied by a person or an object

Table 5.
Smart library seat occupancy and reservation systems methodologies and future research.

6. Conclusions

In this chapter, a systematic literature review was conducted to find out what other researchers found about the Seat occupancy and reservation systems. This process incorporated the identification and comparison of different smart library seat occupancy and reservation systems and their weaknesses. It also helped the researcher to discover about the current operations on seat occupancy and reservation systems in libraries and the gaps that still need to be filled. In addition, previous research that describes characteristics of smart library, highlighting the advantages and disadvantages was explored to further understand real context and the methods that can be used to solve the current seat occupancy and reservation challenge was identified.

To achieve this, the researcher reviewed previously researched papers, articles and journals related to IoT-based seat occupancy and reservation systems.

The IoT-based seat occupation and reservation systems are discussed in detail in this Chapter, illustrating various IoT-based seat occupation and reservation systems using different IoT technologies. These IoT methodologies include LBPH face recognition algorithm, Thing Speak channel, Up-squared board, ultrasonic sensor, Thing Speak cloud, SQLite Studio database, Python, RFID, FSR, Raspberry-Pi, Wi-Fi Module, Node MCU, PHP API and capacitance sensor chip. To establish the seat occupancy and rank the options, some approaches used a single procedure, such as the capacitance sensor technology. A combination of methods, including RFID and FSR, were used by other authors to identify the seat occupancy. To find out how many seats were occupied, researchers used a face recognition system, an ultrasonic sensor, and other tools. The gap in the literature is discussed and the shortcomings of these IoT-based reservation and seat occupation systems are highlighted. The adoption of mixed methods enhances performance. Other studies employed RFID and FSR. While FSR detects seat occupancy and transmits the information to the network layer, RFID is responsible for identifying authorized users who made a specific seat reservation.

Conflicts of interest

There are no conflicts of interest for this book chapter.

Author details

Magauwane Reneilwe Maepa* and Michael Nthabiseng Moeti
Tshwane University of Technology, Polokwane, South Africa

*Address all correspondence to: maepamr@tut.ac.za

References

[1] Abuarqoub A, Abusaimeh H, Hammoudeh M, Uliyan D, Abu-Hashem MA, Murad S, et al. A survey on internet of things enabled smart campus applications. In: Proceedings of the International Conference on Future Networks and Distributed Systems. Cambridge, United Kingdom: Association for Computing Machinery; 19-20 Jul 2017. pp. 1-7. DOI: 10.1145/3102304.3109810. ACM ISBN 978-1-4503-4844-7/17/07

[2] Bayani M, Segura A, Alvarado M, Loaiza M. IoT-based library automation and monitoring system: Developing an implementation framework of implementation. E-Ciencias de la Informació́n. 2018;**8**(1):83-100

[3] Daniel OC, Ramsurrun V, Seeam AK. Smart library seat, occupant, and occupancy information system, using pressure and RFID sensors. In: 2019 Conference on Next Generation Computing Applications (NextComp). Mauritius: IEEE; 2019. pp. 1-5. DOI: 10.1109/NEXTCOMP.2019.8883610

[4] Boboc DI, Cebuc ŞC. Internet of things (IoT). Database Systems Journal. 2019;**10**(1)

[5] Mohamed H, Lamia M. Implementing flipped classroom that used an intelligent tutoring system into learning process. Computers & Education. 2018;**124**:62-76

[6] Brian ALA, Arockiam L, Malarchelvi PDSK. An IOT based secured smart library system with NFC based book tracking. International Journal of Emerging Technology in Computer Science & Electronics (IJETCSE). 2014;**11**(5):18-21

[7] Upala M, Wong WK. IoT solution for Smart library using facial recognition. In: IOP Conference Series: Materials Science and Engineering. Vol. 495, No. 1. Sarawak, Malaysia: IOP Publishing; 2019. p. 012030. DOI: 10.1088/1757-899X/495/1/012030

[8] Bibri SE. The Shaping of Ambient Intelligence and the Internet of Things. Springer; 2015

[9] Maepa MR, Moeti MN. IoT-based smart library seat occupancy and reservation system using RFID and FSR technologies for South African universities of technology. In: Proceedings of The International Conference on Artificial Intelligence and its Applications. Mauritius: Virtual Event, ACM; 9-10 Dec 2021. pp. 1-8. DOI: 10.1145/3487923.3487933. ISBN 978-1-4503-8575-6/21/12

[10] Liang X. Internet of things and its applications in libraries: A literature review. Library Hi Tech. 2018;**38**(1):67-77

[11] Jayawardena C, Reyal S, Kekirideniya KR, Wijayawardhana GHT, Rupasinghe DGIU, Lakranda SYRM. Artificial intelligence based Smart library management system. In: 2021 6th IEEE International Conference on Recent Advances and Innovations in Engineering (ICRAIE). Malaysia: IEEE; 2021. pp. 1-6. DOI: 10.1109/ICRAIE52900.2021.9703998

[12] Goyal P, Sahoo AK, Sharma TK, Singh PK. Internet of things: Applications, security, and privacy: A survey. Materials Today: Proceedings. 2021;**34**:752-759

[13] Pawar N, Bourgeau T, Chaouchi H. Study of iot architecture and application invariant functionalities. In: 2021 IFIP/IEEE International Symposium on

Integrated Network Management (IM). Bordeaux, France: IEEE; 2021. pp. 667-671

[14] Jha K, Anumotu S, Soni K. Security issues and architecture of IOT. In: 2021 International Conference on Artificial Intelligence and Smart Systems (ICAIS). Coimbatore, India: IEEE; 2021. pp. 1381-1385. DOI: 10.1109/ICAIS50930.2021.9395962

[15] Maraveas C, Piromalis D, Arvanitis KG, Bartzanas T, Loukatos D. Applications of IoT for optimized greenhouse environment and resources management. Computers and Electronics in Agriculture. 2022;**198**:106993

[16] Alam I, Sharif K, Li F, Latif Z, Karim MM, Biswas S, et al. A survey of network virtualization techniques for internet of things using sdn and nfv. ACM Computing Surveys (CSUR). 2020;**53**(2):1-40

[17] Bansal S, Kumar D. IoT ecosystem: A survey on devices, gateways, operating systems, middleware, and communication. International Journal of Wireless Information Networks. 2020;**27**(3):340-364

[18] Albishi S, Soh B, Ullah A, Algarni F. Challenges and solutions for applications and technologies in the Internet of Things. Procedia Computer Science. 2017;**124**:608-614

[19] Berek, L. Smart devices and services in the library-the importance of smart libraries. Transactions on Advanced Research. 2021;**17**(2):8-12

[20] Li Y. Application process algorithm analysis of RFID technology. In: In 2021 7th International Symposium on Mechatronics and Industrial Informatics (ISMII). IEEE; 2021. pp. 202-205

[21] Misra G. Kumar V, Agarwal A. Agarwal K, Internet of things (IoT)–a technological analysis and survey on vision, concepts, challenges, innovation directions, technologies, and applications (an upcoming or future generation computer communication system technology). American Journal of Electrical and Electronic Engineering, 2016;**4**(1). pp. 23-32

[22] Sundaravadivazhagan B, Subashini B. A review on internet of things (IoT): Security challenges, Issues and the countermeasures approaches. Psychology and Education Journal. 2021;**58**(2):6544-6560

[23] Igbinovia MO. Internet of things in libraries and focus on its adoption in developing countries. Library Hi Tech News. 2021;**38**(4):13-17

[24] Mani Z, Chouk I. Consumer resistance to innovation in services: Challenges and barriers in the internet of things era. Journal of Product Innovation Management. 2018;**35**(5):780-807

[25] Salih KOM, Rashid TA, Radovanovic D, Bacanin N. A comprehensive survey on the internet of things with the industrial marketplace. Sensors. 2022;**22**(3):730

[26] Whaiduzzaman M, Barros A, Chanda M, Barman S, Sultana T, Rahman MS, et al. A review of emerging technologies for IoT-based smart cities. Sensors. 2022;**22**(23):9271

[27] Zhong RY, Xu X, Klotz E, Newman ST. Intelligent manufacturing in the context of industry 4.0: A review. Engineering. 2017;**3**(5):616-630

[28] Patel KK, Patel SM, Scholar P. Internet of things-IoT: Definition, characteristics, architecture, enabling technologies, application & future

challenges. International Journal of Engineering Science and Computing. 2016;**6**(5):6122-6131

[29] Möller DP. Introduction to the internet of things. In: Guide to Computing Fundamentals in Cyber-Physical Systems. Cham: Springer; 2016. pp. 141-184

[30] Torres A, Paul G. Reaching maximum occupancy: What the numbers tell us about space and ways to improve services. Journal of Access Services. 2019;**16**(2-3):78-93

[31] Hoang, Huy NH, Hettiarachchi G, Lee Y, Krishna Balan R. Small scale deployment of seat occupancy detectors. In: Proceedings of the 3rd International on Workshop on Physical Analytics. Singapore: ACM; Jun 2016. pp. 25-30. DOI: 10.1145/2935651.2935660. ISBN 978-1-4503-4328-2/16/06

[32] Yahaya N, Sangaram D, Mohammed Ali S, Tay A. SMU libraries: "Sensory" in library spaces. Singapore Journal of Library and Information Management. 2017;**45**:61

[33] Rhee E. RFID-based cultural performance progress management system. International Journal of Internet Technology and Secured Transactions. 2022;**12**(6):503-516

[34] Hirekhan SN. Application of RFID and biometric Technology in Libraries: A study. In: Knowledge Management in Higher Education Institutions. 1st ed. Rajasthan, India: Manipal University Jaipur; 2022. p. 159. Available from: www.jaipur.manipal.edu

[35] Igwe KN, Sulyman AS. Smart libraries: Changing the paradigms of library services. Business Information Review. 2022;**39**(4):147-152

[36] Jadhav CJ, Jadhav SS, Sancheti VM, Hajare SS. Smart library management system using RFID technology. International Research Journal of Engineering and Technology (IRJET). 2017;**4**(5):2831

[37] Massis B. The internet of things and its impact on the library. New Library World. 2016;**117**(3/4):289-292

[38] Siddaway AP, Wood AM, Hedges LV. How to do a systematic review: A best practice guide for conducting and reporting narrative reviews, meta-analyses, and meta-syntheses. Annual Review of Psychology. 2019;**70**:747-770

[39] Kitchenham BA. Systematic review in software engineering: Where we are and where we should be going. In: Proceedings of the 2nd International Workshop on Evidential Assessment of Software Technologies. Lund, Sweden; 22 Sept 2012. pp. 1-2. ACM 978-1-4503-1509-8//12/09

[40] Dabić M, Vlačić B, Paul J, Dana LP, Sahasranamam S, Glinka B. Immigrant entrepreneurship: A review and research agenda. Journal of Business Research. 2020;**113**:25-38

[41] Paul J, Lim WM, O'Cass A, Hao AW, Bresciani S. Scientific procedures and rationales for systematic literature reviews (SPAR-4-SLR). International Journal of Consumer Studies. 2021;**45**(4):1-16

[42] Paul J, Benito GR. A review of research on outward foreign direct investment from emerging countries, including China: What do we know, how do we know and where should we be heading? Asia Pacific Business Review. 2018;**24**(1):90-115

[43] Kamboj S, Rahman Z. Marketing capabilities and firm performance: Literature review and future research agenda. International Journal of

Productivity and Performance Management. 2015;**64**(8):1041-1067

[44] Moher D, Liberati A, Tetzlaff J, Altman DG, PRISMA Group. Preferred reporting items for systematic reviews and meta-analyses: The PRISMA statement. Annals of Internal Medicine. 2009;**151**(4):264-269

[45] Khatoon S, Rehman V. Negative emotions in consumer brand relationship: A review and future research agenda. International Journal of Consumer Studies. 2021;**45**(4):719-749

[46] Tranfield D, Denyer D, Smart P. Towards a methodology for developing evidence-informed management knowledge by means of systematic review. British Journal of Management. 2003;**14**(3):207-222

[47] Boell SK, Cecez-Kecmanovic D. On being 'systematic' in literature reviews. In: Willcocks LP, Sauer C, Lacity MC. editors. Formulating Research Methods for Information Systems. Vol. 2. London: Palgrave Macmillan; 2015. pp. 48-78. DOI: 10.1057/9781137509888_3

[48] Watson R. Quantitative research. Nursing Standard: official newspaper of the Royal College of Nursing.. 2015;**29**(31):44-48. DOI: 10.7748/ns.29.31.44.e8681

[49] McLean R, Antony J. Why continuous improvement initiatives fail in manufacturing environments? A systematic review of the evidence. International Journal of Productivity and Performance Management. 2014;**63**(3):370-376

[50] Zorzela L, Golder S, Liu Y, Pilkington K, Hartling L, Joffe A, et al. Quality of reporting in systematic reviews of adverse events: Systematic review. BMJ. 2014;**348**:f7668

[51] Wohlin C. Guidelines for snowballing in systematic literature studies and a replication in software engineering. In: Proceedings of the 18th international conference on evaluation and assessment in software engineering. London, England: BC United Kingdom Copyright; 13 - 14 May 2014. pp. 1-10. DOI: 10.1145/2601248.2601268. ACM 978-1-4503-2476-2/14/05

[52] Wang Y, Wei Y. Research on library functional layout based on intelligent occupying system. In: Journal of Physics: Conference Series. Vol. 1176, No. 3. IOP Publishing Ltd; 2019. p. 032010. DOI: 10.1088/1742-6596/1176/3/032010

[53] Zhou D. Case study on seat Management of University Library Based on WeChat public number client—Taking Jianghan university library as an example. In: 2019 4th International Conference on Mechanical, Control and Computer Engineering (ICMCCE). Hohhot, China: IEEE; 2019. pp. 630-6303. DOI: 10.1109/ICMCCE48743.2019.00146

[54] Xu X, Guo ZM, Zhu YR. Design of seat management system in Smart library. In: 2022 10th International Conference on Orange Technology (ICOT). Shanghai, China: IEEE; 2022. pp. 1-3. DOI: 10.1109/ICOT56925.2022.10008128

[55] Liu Y, Ye H, Sun H. Mobile phone library service: Seat management system based on WeChat. Library Management. 2021;**42**(6/7):421-435

[56] Lipat J, Rabano CE, Mamauag MA, Galang MI, Sparse CJ. A Reservation and Computer Vision-Based Room Occupancy System for Malayan. Bordeaux, France: Colleges Laguna's Center for Learning and Information Resources; 2022. pp. 667-671

[57] Yang S, Wang X. Development of 3D library application system. Journal of Simulation. 2020;**8**(3):85

[58] Jing Y, Mao L, Xu J. Research on a library seat management system. In: 2019 2nd International Conference on Information Systems and Computer Aided Education (ICISCAE). Dalian, China: IEEE; 2019. pp. 297-301. DOI: 10.1109/ICISCAE48440.2019.221639

[59] García-Granja MJ, Blázquez-Parra EB, Cimadomo G, Guzmán-Navarro F. Development of an innovative seat reservation system for university buildings based on BIM technology. Buildings. 2022;**12**(11):1786

Chapter 2

Methods for Detection and Prevention of Vulnerabilities in the IoT (Internet of Things) Systems

Vesna Antoska Knights and Zoran Gacovski

Abstract

In this chapter, the problem of detection and prevention systems for Internet of Things attacks is discussed. We start with the term Internet of Things (IoT) that defines the use of intelligently connected devices and systems for data collection *via* embedded sensors and actuators in physical devices. IoT is omnipresent today and is expected to expand globally in the years ahead. This type of progress will provide services that improve the quality of human life and the productivity of enterprises, while creating the possibility for the so-called "Connected Life." The goal is to achieve protection against intruders who break into IoT systems to obtain certain sensitive data, gain control, or commit any kind of abuse. Reliability, integrity, and availability represent three different aspects in the field of security that should be achieved in systems such as the Internet of Things. In this work, we present three major areas that can help to mitigate the security risks in IoT systems. Also—two methods for intrusion detection are elaborated—signature-based and anomaly-based models. In the last section of this chapter, we present a real-world example that has already been implemented in reality (Intrusion detection system based on Snort at eggs/poultry farm).

Keywords: Internet of Things, vulnerabilities detection, prevention, threats, signature-based IDS, system security

1. Introduction

Internet of Things is a widespread, intelligent network composed of smart devices that enables the implementation of advanced services in housing, manufacturing, transport, health, and other sectors, as well as enabling a new ecosystem for application development. Examples of widespread IoT systems include the following:

- Fixed devices (home appliances, surveillance, smart meters, smart grid, street lamps).

- Mobile devices (in-car monitoring, transport/logistics, mobile robots, etc.).

- Personal devices (remote health monitoring, assisted living wristband, pharma sensors).

IoT devices share the sensor data they collect by connecting to an IoT gateway or other edge device—from where the data is sent to the cloud to be stored and analyzed. The data can also be analyzed locally on edge devices—in real time.

Today, with more connected "things" than the population of the Earth, the issue of IoT security is a major challenge [1]. More billions of IoT devices increases the threat and opens up the possibility of numerous attacks on the devices themselves. To address these security challenges, it is essential to explore frameworks like the 'Security Framework for Internet of Things proposed by El-Gendy and Azer [2]. Hacked devices cause disruption of connectivity and can also serve as a starting point for attacks on other devices and systems. The issue of confidentiality, availability, and integrity of data is greater than ever; therefore, it is necessary to ensure the functionality of CIA (Confidentiality, Integrity, Availability) through encryption and other protection methods.

Many authors have researched the appearing attacks and risks of the IoT devices:

- An attack on medical devices and equipment that the attacker has "locked" with a malicious program and demands a ransom (ransomware) [3]. One of the insecure medical devices that is in direct contact with patients and provides them with "life" is the cardiac electrostimulation (or "pacemaker"), with the associated programming device, due to the lack of passwords or any authentication.

- An attacker takes full control of the vehicle while driving [4]. Although the automotive industry invests a lot in product safety, research and testing have revealed vulnerabilities in numerous sensors that are present in newer cars—such as brake activation, vehicle steering, and other controls—for example,. in Tesla model S.

- The oil industry, as one of the critical infrastructures on which many other industries depend, uses numerous IoT sensors, devices, and applications in its business processes for the purpose of monitoring and control [5]. Unsafe IoT sensors and devices are present in the operation of the pipeline that connects the oil and gas fields and the refineries. Critical infrastructure is also at risk and is a frequent target of attacks.

Many IoT vulnerabilities could be mitigated with recognized security best practices, but too many products today do not include even basic security measures. There are many factors that contribute to this lack of security. It is unclear who is responsible for security decisions in a situation where one company designs the device, another supplies the component software, a third manages the network in which the device is embedded, and a fourth uses the device. This challenge is compounded by the lack of comprehensive, widely adopted international regulations—norms and standards for IoT security. Addressing the growing concerns about IoT security, it is vital to consider existing security protocols as highlighted in Maamar et al.'s comprehensive survey of Internet of Things security protocols [6]. Moreover, exploring recent surveys, such as the work by Alaba et al. [7], can provide valuable insights into the overall landscape of Internet of Things security.

Given the numerous incidents in the past few years that exploited IoT devices and ecosystems, countries have begun to recognize and be aware of the problem. In order for the IoT to function in a secure environment in the future, they started to formally standardize and legally regulate the Internet of Things and adhere to the best security practices.

However, advancements in robotics, such as the integration of anthropometric robots, offer potential solutions to enhance IoT security. For example, the use of mobile anthropometric robots with their ability to interact with the IoT infrastructure can improve security measures by enabling monitoring, threat detection, and response capabilities in IoT environments [8].

In addition to security protocols and surveys, it is essential to recognize the role of intrusion detection systems (IDS) in safeguarding IoT environments. Gupta et al. conducted a survey focused on intrusion detection systems in wireless sensor networks [9], shedding light on their significance in enhancing security in IoT ecosystems.

In our research, we have implemented and tested an intrusion detection system (IDS) at an egg/poultry farm Vezeshari. Our IDS was based on Snort—open-source signature-based intrusion detection system. We have set up the system rules to detect four types of IoT attacks: dynamic login attempts, XML injection attacks, SQL injection attacks, and Firmware (command) injection attacks. All of these attacks utilized code injections that target the wireless layer of the IoT system. The wireless frames from different Wi-Fi components of the IoT system are prone to these attacks, so we tested injection attacks at these points. To test our IDS, we invited ethical hackers from multiple countries that conducted orchestrated attack attempts. In most of the cases (over 80%), our IDS was able to detect and alert on the attack attempts—and the results are presented in Section 5. Our IDS is signature-based and to overcome its limitations, we plan to follow Tacker's et al. approach [10] that implements an anomaly-based IDS based on machine learning, and it will be our future research.

This chapter is organized as follows: in Section 2, we elaborate on the vulnerability risks of IoT systems; in Section 3, we present risk prevention and management for the IoT systems; in Section 4, we explain the usage of intrusion detection and prevention systems (IDS/IPS). In Section 5, a real-world use case (Vezeshari's IDS) is presented with methodology and key results, and then in Section 6, we give final conclusions.

2. Vulnerability risks for IoT systems

Anything that can disrupt the operation, integrity, and availability of an IoT device, or network of IoT devices is a threat. There are different types of threats. There are natural threats, such as floods, earthquakes, or storms. There are unintentional threats that result from accidents or mistakes. Finally, there are deliberate threats that result from ulterior motives. Each of these types of threats can be fatal to the IoT infrastructure.

The IoT risks are divided into four categories (**Figure 1**).

- Base risk—the possibility of device breach (data loss, physical harm, integrity risk).
- Harm to other stakeholders—including those not previously anticipated.
- Risk for future misuse cases.
- Future aggregation risks—network effect on other connected devices.

Therefore, the general security requirements for IoT systems (**Figure 2**) must include the following:

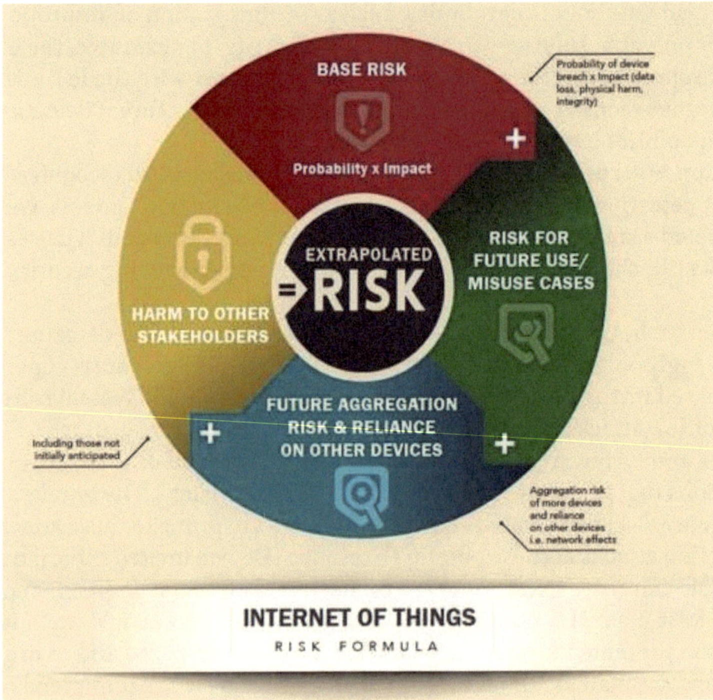

Figure 1.
IoT risks.

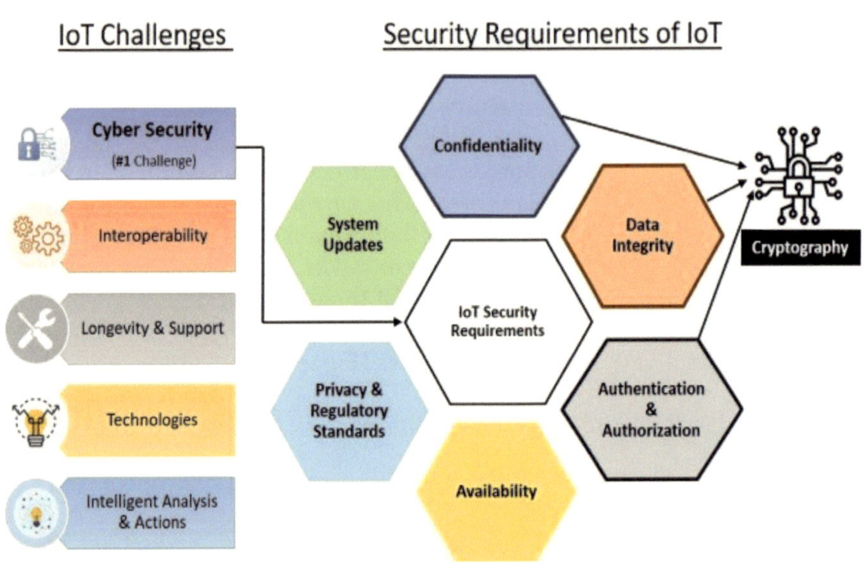

Figure 2.
IoT security challenges and requirements.

- Availability—the assurance that the services and resources (IoT devices) will be available and usable when requested by the user.

- Authentication—only authorized users have access to devices and services based on the following controls: authorization (granting and revoking of access rights), delegation (transfer of part of the rights of one entity to another), and authentication of users (identity check).

- Confidentiality—protects the existence of the connection, the flow of traffic, and the content of the information from disclosure to unauthorized users [11].

- Restoration—assurance that information and IoT devices will survive an attack and that availability can continue after the attack.

IoT vulnerability is an integral weakness of the design, configuration, or implementation of the IoT system that weakens protection. Most vulnerabilities can be found in one of these sources:

- *Bad design*: Hardware and software systems that contain design holes that can be misused. Basically, IoT systems are created with security holes. An example of this is the malware mail attack that leads to gaining access to IoT devices. These vulnerabilities (holes) can occur in any environment—Windows, Linux, Android, which may be resolved with a service pack, etc.

- *Poor implementation*: IoT systems that are incorrectly configured are susceptible to attack. This type of vulnerability usually results from inexperience, insufficient training, or irresponsible work. An example of this type of vulnerability would be a system that has no privileges for restricted access to critical devices and thus allows access to the devices.

- *Poor management*: Inadequate procedures and insufficient checks. Security measures cannot operate in a vacuum; they must be documented and monitored. Even the simplest things like daily system backup must be verified. There is a need to allocate responsibility for IoT devices and share responsibility for others. For this issue, the IoT provider should ensure that procedures are followed and that no person has total control over the system.

Although there are only three types of vulnerabilities, they can manifest in many ways. The first security rule is the physical protection of devices and networks. Central hosts and servers should be stored in separate rooms that can only be accessed by authorized personnel (owners). Routers, communications equipment, and portable media (disks, smartcards) should also be stored in secure locations with limited access. As part of this process, individuals and companies must consider the physical and natural environment in which IoT operates. The possibility of earthquakes, fires, floods, and other unforeseen accidents should be considered and properly planned. Accordingly, owners must ensure the security of all media (disks, tapes, smartcards) that contain vital information and make regular data backups.

Communication is the transmission of information through a medium. As such, it is inevitably vulnerable to interception, monitoring, burglary, etc. Owners should also take care of other forms of communication interception (Wi-Fi, antennas, etc.). Network and packet eavesdroppers are common tools that can read network flow.

It is important to note that every network and IoT system has vulnerabilities. Human mistakes, carelessness, laziness, greed, and rage pose the greatest threat to infrastructure with possible high damage. Moreover, human vulnerability and the risks associated with them are the most difficult to defend.

3. Risk prevention and management for the IoT systems

The IoT risk sources can be classified into four categories, as displayed in the following table (**Table 1**).

IoT risks can be treated (mitigated) *via*:

- Avoidance—means avoiding the risk by eliminating the risky process or resources by modifying the process.
- Mitigation—means mitigating risk by implementing measures to reduce risk, for example by improving existing security measures and controls.

Domain	Risk source
Sensor layer	• Physical characteristics—small dimensions require even smaller components with limited (security) capabilities. • Device price—cheaper device price means cheap components without security features embedded. • Power consumption—the long interval of usage requires energy-efficient components that do not possess security capabilities. • Wireless communication—this enables interception of the signal, and if the data is not encrypted, it is subject to intrusion, theft, and misuse. • Heterogeneity—a lot of different standards makes it difficult to provide good security.
Access layer	• Wireless technology—wireless data transmission opens up the possibility of unauthorized interception and analysis of traffic. • Traffic convergence between multiple devices/users in one node—connection of a large number of devices in one point (switch/hub) can be misused in a large number of attacks (traffic eavesdropping, MitM, DoS, etc.)
Network layer	• Traffic routing—OSPF, BGP, and other traffic routing algorithms have flaws that can be exploited to compromise security. • Public routers—they can be subject to attacks like DDoS.
Application layer	• A large number of penetrations—one cloud server manages the data of a large number of private and business users, which raises the issue of data segmentation, privacy, confidentiality, and the like. • Immature technology—the rapid development of services based on cloud computing raises the level of risk due to the insufficient research on security flaws and protection methods. • Enrolling many users in one physical server—identified security flaws in virtualization, the exploitation of which can cause harm to a large number of users at the same time.

Table 1.
IoT risk sources.

- Transfer—means the transfer of the consequences of the harmful effect of the risk to other natural or legal persons. For example, by purchasing an insurance policy against a harmful event or by agreeing on the compensation that the service provider would be obliged to pay for certain harmful events in the case of outsourcing the process.

- Acceptance—implies acceptance of the potential consequences of the harmful effect of the risk. The organization is aware of the risk, but the conclusion is that the costs of procurement and annual maintenance of the security system are greater than the potential lost income and losses caused by a damaged reputation, and it has decided to accept the risk without implementing additional measures.

The best practices for IoT security are defined by the IEEE—the world's largest professional organization for technology advancement. In February 2017,—IEEE issued the document "Internet of Things (IoT) Security Best Practices" [12]. The document is divided into three areas, with recommendations for each area:

- Device protection with prescribed measures and recommendations:
    - Make the hardware resistant to unauthorized use.
    - Provide regular firmware updates and upgrades.
    - Conduct dynamic testing.
    - Prescribe procedures for data protection on device disposal.
- Network protection with the following recommendations:
    - Use strong authentication.
    - Use strong encryption and security protocols [13].
    - Minimize device throughput.
    - Segment the network.
- Protection of the entire IoT system with the following recommendations:
    - Protect sensitive data.
    - Promote and conduct ethical hacking.
    - Standardization of devices and certification of personnel and organizations.

The conclusion is that the mentioned recommendations and measures should be used by manufacturers who produce IoT devices, by programmers and engineers who come up with the design of devices and systems, by researchers and testers to evaluate IoT systems, and by legislators when creating security and other acts that cover the IoT area.

4. Intrusion, detection, and prevention systems

The main goal of the intrusion detection and prevention system is to prevent situations that are not categorized as normal, but as suspicious (caused by the misuse of information), and to detect attacks and achieve security when such modes occur [14–16]. It also includes documentation of existing threats and serves as a controller of the IoT system security design. Intrusion Detection System (IDS) provides attack information, advanced diagnosis, systems recovery, and various investigations that allow ongoing events to be performed, including stopping the attack, terminating the network connection or user session, blocking the availability of the attack target, and changing the appropriate security.

The IDS [16], upon detection of an intrusion into the IoT system, raises an audio or video alarm, or sends a warning in the form of an e-mail message or text message to a smartphone. An improvement of this technology is the System (IPS), which can detect an intrusion and, furthermore, prevent that intrusion from Intrusion Prevention being successful through an appropriate active response.

The components of the IDS include (**Figure 3**) the following:

- Sensor (agent)—the agent is the module that collects event data and analyzes system activities. In the case of using an IDS, the agent is called a "sensor".

- Management server—it is responsible for analyzing information received from ongoing activities and deciding whether there is an attack in progress. This server also uses information from other elements, e.g. signatures and profiles, to complete the analysis.

- Management interface (console)—serves as an interface between the IDS and the administrator. The console is used to monitor system events received by the IDS. Some consoles are also used to configure agents and perform software upgrades.

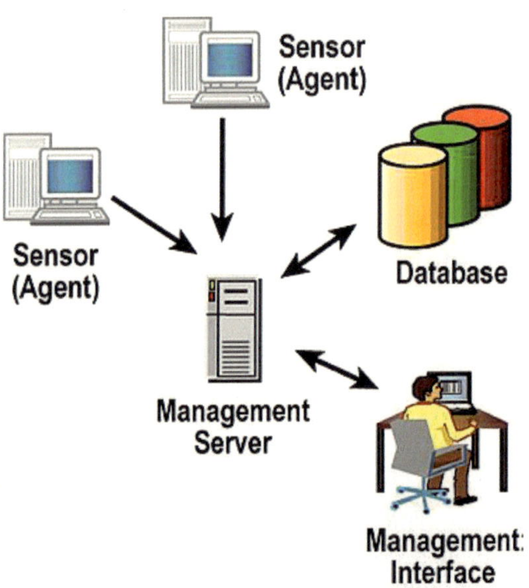

Figure 3.
Components of the IDS system.

- Database—the database server is used to store the information received from the agents and the management server. The management server also uses the database server to complete the analysis.

There are two major methods for intrusion detection—a signature-based model and an anomaly-based detection model.

Signature-based detection model—this method works by comparing current events with certain known signatures. Below are some examples of signatures:

- An attempt to log into the system as an administrator (root), which violates the system's security policy.

- Email messages with the subject "Free Screensavers" and containing an attachment in the message "screensaver.exe", which refers to spam messages.

- The operational system log for entering the system with code 645, which refers to the server is disabled and should be audited.

This type of IDS/IPS is very effective due to its low complexity in the implementation and detection process. It simply compares the current activity with the stored signatures to find any pattern in order to detect the attack. In addition, this model produces very specific attack reports compared to the anomaly-based model that is described below. A disadvantage of the signature-based detection and prevention model is its inability to detect new unknown attacks because the system has no signatures to enter the system for new attacks.

Anomaly-based detection model—this model detects attacks based on profiles. Profiles contain the pattern or normal behavior mode in which the system is used. Profiles are derived based on specific users, networks, or applications. They are created by monitoring system usage over a period of time, known as the evaluation period (**Figure 4**). This model compares current activities with profiles in order to detect abnormal activity, which in most cases indicates a seizure.

Since system and network usage are not static and always contain some variation over time, the profile must also adjust accordingly over time. Therefore, after creating profiles during the evaluation period, the detection and prevention system changes the profiles over time. Below are examples of profiles:

- User profile contains 5% email activity. When the detection and prevention system using the anomaly-based model detects that the email activity in the system is more than 5%, it will consider it as an attack.

- Over the course of a few weeks, the average user opens, reads, and writes to the file system 2% of the time. When the detection and prevention system detects a sudden increase in file system activity, it will consider it an attack.

The advantage of the anomaly-based model is that it can detect even unknown attacks by comparing current abnormal events with events considered normal. Furthermore, this model can also be more efficient than a signature-based model, since there are a large number of signatures to compare when using the same model. On the other hand, the incidents detected by the anomaly-based model are not very specific and therefore require additional effort on the part of the administrator to determine the root point of the attack. In addition, this model is subject to the

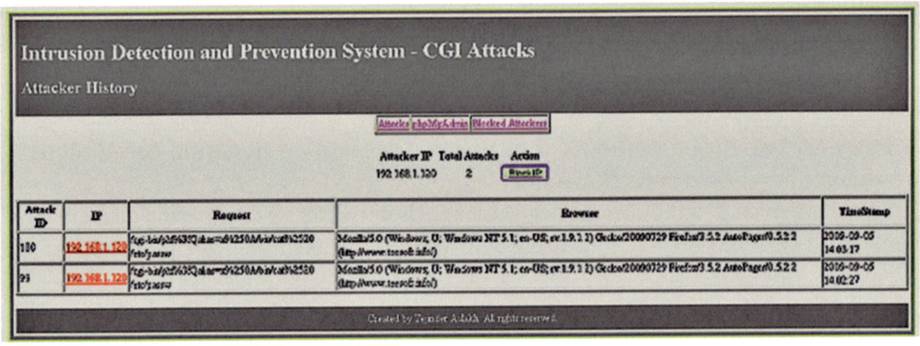

Figure 4.
Attacks registered by an IDS/IPS.

so-called "slow attack". In this type of attack, the attacker first learns the threshold between normal and abnormal activity in the system. The attacker will then slowly attack the system to ensure that the activities during the attack do not reach a threshold that will result in the detection of the attack by the anomaly-based model.

5. Real-world use case: methodology and results

Vezeshari is a poultry and eggs farm from Zhelino, North Macedonia. They have chosen Xively (software by LogMeIn) for their needs and working processes, as well as their Xively IoT platform. By these, they can monitor the internal air temperature, air humidity, CO_2 amount, and water nutrient level, and also –get a history of security alerts (**Figure 5**). Security, scalability, and expertise are important deciding factors, and by choosing an IoT solution, Vezeshari can now handle the entire network of connections. They also have overview and support for different scenarios, while getting the best guidance and advice from professionals (Xively).

By getting secure, reliable, and fast access, Vezeshari maintenance team can now excel in other areas of the farm and overall business:

- Access to real-time data allows Vezeshari to quickly diagnose problems and take immediate action to resolve them. When working with livestock, real-time information on climate conditions is essential, as the environment needs to be optimal at all times.

- Vezeshari customers can gain insight into the current state of their farm to ensure the process is running smoothly. This reduces the total time the farmer needs to work on the farm.

- The developed Vezeshari analyzer provides better insight into how people use their products and how they would like the products to be used. This helps Vezeshari refine and optimize its products to meet all user needs.

- Through the mobile Connect App, farmers can set parameters for optimal growing conditions and receive notifications and alerts related to parameters in the farm. Trusted software provides a better customer experience and enables instant problem resolution.

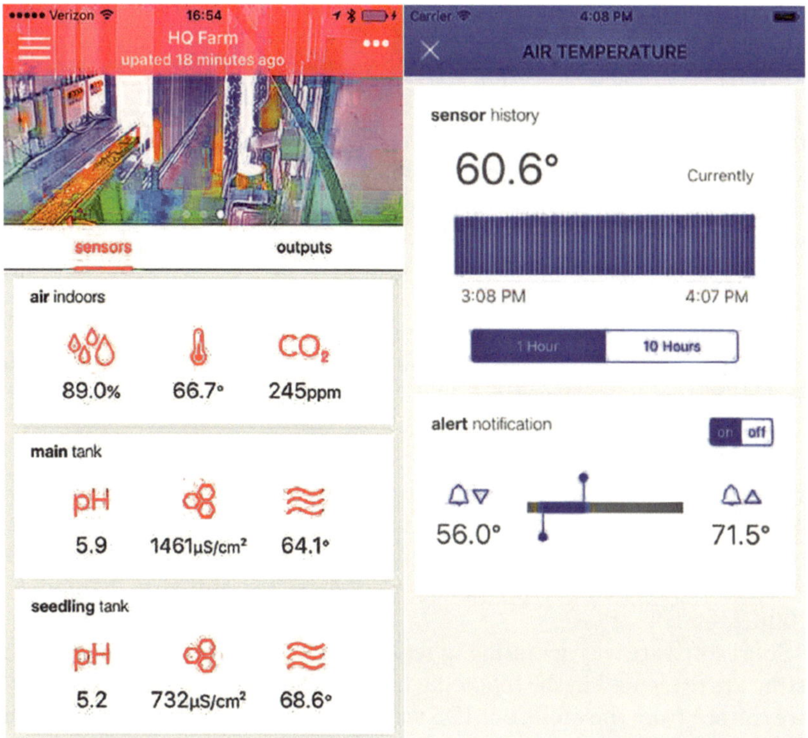

Figure 5.
Chicken farm application—allows monitoring of internal air temperature, air humidity, CO_2 amount, water nutrient level, as well as a history of alerts (presented graphically).

- Vezeshari employees have insight into day-to-day farm activities that gives them the ability to proactively meet customer needs, while also growing their business by selling consumables that are needed quickly based on farm usage.

Today, the entire Vezeshari working process is supported by state-of-the-art systems for detection and prevention of attacks that are updated and optimized daily. Thus, they achieve protection that is reliable even against the biggest threats and attacks in the system. Attack detection and prevention systems are composed of various antivirus and antispyware software, as well as the latest versions of firewalls for protection. Other methods using an increased level of security and data encryption, such as protection of the smartphone and tablet applications themselves through various types of generated passwords, are used in order to enable the detection and prevention systems to successfully recognize and prevent attacks in the systems.

To test the resilience of Vezeshari's network, we have implemented signature-based IDS (intrusion detection system). We used Snort—open-source signature-based IDS—and we have set up its rules to detect the following types of attacks:

- Dynamic login attempts.
- XML injection attacks.

- SQL injection attacks.
- Firmware (command) injection attacks.

To test our IDS, we invited ethical hackers from several countries (Hungary, Israel, France, USA, Netherlands) *via* forums and mailing lists (total of 56 participants). The attackers utilized code injections that targeted the wireless layer of the IoT system. The wireless frames from different Wi-Fi components of the IoT system are prone to these attacks—so we tested injection attacks at these points. To test our IDS, the ethical hackers conducted orchestrated attack attempts from multiple locations.

We've installed and configured Snort software in accordance with [17]. We opened the extracted folder *snortrules* then navigated to the *etc* folder and copied the snort.conf file then pasted it in *C:Snortetc* folder. Next we moved the folders so_rules, preproc_rules to the *C:Snort* directory path.

We entered and enabled the rules set before launching Snort, that is, we enabled ICMP rules so that Snort can detect any ping probes to the system while running. For example, a rule to detect a suspicious TCP intrusion (access rule) is in the form:

alert tcp any any -> any any (msg:"Suspicious User-Agent detected"; flow:to_server,established; content:"User-Agent|3a| "; nocase; content:"curl|2f|";nocase; sid:1000002; rev:1;)

The Snort software was operating in real time and results of the intrusion detection testing are presented in the following table (**Table 2**).

As we can see from the table, our IDS was able to successfully detect and alert on the attack attempts in most of the cases (over 80%). The most successful our system was for the firmware command injection attacks (95.3%), and the least successful in case of XML injection attacks. The rate of success varies for different attacks, because of the different capability to hide the harmful code within the legitimate IP packets. (The Snort rules can easily detect a harmful code within the firmware commands, rather than within XML code.)

The results indicate that our IDS for detection and prevention of attacks should be updated and optimized weekly, in order to achieve reliable and robust protection even against the most current threats and attacks, especially for this new IoT technology.

Of course our approach (usage of signature-based detection and Snort) has potential limitations, such as the following:

- Potential false positives (false negatives) reported by the IDS. If we applied the default Snort configuration, it would report a lot of false alarms.
- Interfaces that have overlapping IP addresses as matching criteria in Access rules might not be detected by Snort rules as intended.

Attack type	Signature	Total attempts	Detection/alert rate
Dynamic login attempts	20,790,001	348	311 (89%)
XML injection attacks	20,810,001	1250	1009 (80.7%)
SQL injection attacks	1,220,002	282	240 (85.1%)
Firmware (command) injection	1,310,001	455	434 (95.3%)

Table 2.
Detection rate of four different types of attacks.

- Services based on the payload of connections, such as network applications, URL categories, or URL list applications, can hardly be implemented as traffic Access rules for Snort inspection.

- Snort inspection is not supported for Virtual engines (WAN access).

To overcome these limitations we plan to extend our IDS by implementing an anomaly-based detection as the one elaborated in [10] by Tacker et al., and we are discussing this possibility in the last section of this paper (future work).

6. Conclusion and future work

The Internet of Things, which is also called the fourth industrial revolution, truly deserves that name. The tendency of IoT is to connect all disconnected devices. Today, life without the Internet of Things is unimaginable. Numerous devices from households, (air conditioners, lighting, video surveillance) to numerous devices in practically all industries (agriculture, healthcare, finance, energy, the automotive industry etc.), in order to improve the quality of life and increase economic growth, are interconnected, integrated and share data in real-time with the help of networks of all networks, the Internet.

But there is also a real danger that these devices will be exposed to cyber-attacks and that the confidentiality, integrity, and availability of data will come into question. IoT devices generate a large amount of data; therefore the question of the security of these devices and the entire IoT ecosystem arises, as well as the question of privacy.

Of course, as for everything else, it is necessary to carry out a risk assessment, from the identification of the resource, the vulnerability of the resource itself, possible threats that can exploit the vulnerability, and in the case of an attack, to determine the consequences that the attack could cause. The practice has shown the vulnerability of IoT devices due to several reasons, from the very physical characteristics of the devices, which are small in size, mostly untested for safety before use, low price, and low energy consumption. It is this sensor layer that is most at risk, in contrast to the access, network and application layers, where the risk is mostly assessed as low to medium.

Also, IoT technologies are not followed by legislation either, only in the last two to three years have societies recognized the problem and become aware of the risks and started the process of establishing a legislative and standardization framework that will regulate the IoT area.

The only way to stop attacks is to know the techniques used to attack. Therefore, organizations' security systems will need to adopt the most robust model or mechanism that provides the strongest protection against threats to ensure that the system remains secure. The attack detection and prevention system (IDS/IPS) ensures that attacks are detected and prevented using multiple approaches. Active attack detection and prevention systems aim to limit the damage that attackers can cause by building a local network that is resistant to the appropriate attack or threat.

In the last section of the paper, we presented a real-world use case (a poultry farm) that has acquired an IoT solution (Xively) and is now able to monitor the internal air temperature, air humidity, CO2 amount, water nutrient level, and get a history of security alerts that now can be controlled. At Vezeshari, we have implemented an open-source IDS in Snort to detect four types of IoT attacks: dynamic login attempts, XML injection attacks, SQL injection attacks, and Firmware (command) injection attacks. All of these attacks utilize code injections that target the wireless layer of the IoT system. The wireless frames from different Wi-Fi components of the IoT system are prone to these attacks so we tested injection attacks at these points.

To test our IDS we invited ethical hackers from multiple countries that conducted orchestrated attack attempts. In most of the cases (over 80%) our IDS was able to detect and alert on the attack attempts, and the results are presented in Section 5. The rate of success varies for different attacks, because of the different capabilities to hide the harmful code within the legitimate IP packets. Therefore, our system for detection and prevention of attacks should be updated and optimized weekly, to achieve reliable and robust protection even against the biggest threats and attacks in the system, especially in a new branch of technology such as the Internet of Things.

In future work, we plan to extend our IDS by implementing anomaly-based detection, i.e. by utilizing machine learning techniques. Anomaly-based IDS-s require larger processing resources, but they are superior in the detection of new, previously unknown threats. They are also adaptive and dynamic, as they can learn from the network behavior and update the baseline accordingly. There are eight methods for detecting traffic anomalies in real-time data, namely: projection-based methods, regression-based methods, support vector machines, decision tree–based methods, density-based methods, clustering, distance-based, and time series–based methods.

We plan to apply support vector machines [18] and decision trees [19]. Both of these machine learning methods are based on training the detector (IDS system), which will learn and be able to detect the real anomalies (intrusions). In the testing phase, a new data set will be used to develop the system's capacity to generalize to previously unseen intrusions. Support Vector Machine (SVM) method is a classification approach where support vectors form the boundaries of a class. In detecting anomalies, a single-class SVM will be used to define a normal class, and points that are outside the class boundaries can be defined as anomalies. Tree-based methods create a tree structure from data, where the tree will be updated with new data, but if new data causes significant changes in the tree structure, the model needs to be re-trained.

The field of IoT security is a contemporary and dynamic field of research. The broader significance of our research is that we implemented a cost-effective IDS in a real poultry farm. Our findings presented in this paper illustrate an IDS that is affordable and easy to implement in many small and medium businesses. We proved that the IoT system should not be put in function without providing its security. Our system is able to detect multiple threats, including DoS (denial of service) and malware (virus attacks). Our ultimate goal will be to develop an autonomous IDS and apply state-of-the-art techniques of machine learning and deep learning that can learn from the big IoT data. In addition, such future IoT IDS would have features such as self-configuration, self-optimization, self-protection, and self-healing.

Author details

Vesna Antoska Knights[1*] and Zoran Gacovski[2]

1 Faculty of Technology and Technical Science—Veles, University "St Kliment Ohridski"—Bitola, Bitola, Republic of North Macedonia

2 Faculty or Faculty of Technical Sciences, Mother Teresa University—Skopje, Skopje, Republic of North Macedonia

*Address all correspondence to: vesna.knights@uklo.edu.mk

References

[1] Ashoor AS, Gore S. Importance of intrusion detection system (IDS). International Journal of Scientific and Engineering Research. 2011;**2**(1):75-78. ISSN: 2229-5518

[2] El-Gendy S, Azer MA. Security framework for internet of things (IoT). In: 2020 15th International Conference on Computer Engineering and Systems (ICCES). Piscataway, USA: IEEE; 2020. pp. 1-6. DOI: 10.1109/ICCES51560.2020.9334589

[3] Newman LH. A new pacemaker hacks put malware directly on the device. 2018. Available from: https://www.wired.com/story/pacemaker-hack-malware-black-hat/

[4] Mahaffey K. Hacking a Tesla Model S: What we found and what we learned? 2015. Available from: https://blog.lookout.com/hacking-a-tesla

[5] Hanes D, Salgueiro G, Grossete P, Barton R, Hanes J. IoT Fundamentals: Networking, Technologies, Protocols and Use Cases for the Internet of Things. San Jose, USA: Cisco Press; 2017

[6] Maamar Z, Yahyaoui H, Mosbah M, Hemery J. Security for Internet of Things: A survey of existing protocols. In: Proceedings of the International Conference on Internet of Things and Big Data. New York, USA: ACM; 2018

[7] Alaba F, Othman M, Hashem IAT, Alotaibi F. Internet of things security: A survey. Journal of Network and Computer Applications. 2017;**88**:10-28

[8] Antoska V, Jovanović K, Petrović VM, Baščarević N, Stankovski M. Balance analysis of the Mobile Anthropomimetic robot under disturbances – ZMP approach. International Journal of Advanced Robotic Systems. 2013;**10**(4):1-10. DOI: 10.5772/56238

[9] Gupta BB, Nayyar A, Rautaray SS. A survey of intrusion detection systems in wireless sensor network. Journal of Network and Computer Applications. 2016;**60**:19-31

[10] Tucker J, Coughlan M, Nelson T, Klimkowski B. Implementing an anomaly-based intrusion detection system: Focus on internal threat – Masquerade attacks. American International Journal of Contemporary Research. 2016;**6**(4):1-11

[11] Abomhara M, Koien G. Security and privacy in the internet of things: Current status and open issues. In: PRISMS 2014 - International Conference on Privacy and Security in Mobile Systems (PRISMS 2014). Aalborg, Denmark, USA: IEEE; May 2014

[12] IEEE – Institute of Electrical and Electronics Engineers. Internet of Things (IoT) Security Best Practices. 2017. Available from: https://standards.ieee.org/wp-content/uploads/import/documents/other/whitepaper-internet-of-things-2017-dh-v1.pdf

[13] Thakor VA, Razzaque MA, Khandaker MRA. Light weight cryptography algorithms for resource-constrained IoT devices: A review, comparison and research opportunities. IEEE Access. 2021;**9**:28177-28193

[14] Bilgin D, Gacovski Z, Pivovarov V, Goracinova L. System for detection of network threats based on classifiers. TEM Journal. 2014;**3**(2):120-126

[15] Sandhu UA, Haider S, Naseer S, Ateeb OU. A survey of intrusion

detection & prevention techniques. In: International Conference on Information Communication and Management IPCSIT. Singapore: IACSIT Press; 2011

[16] Rudner M. Cyber-threats to critical national infrastructure: An intelligence challenge. International Journal of Intelligence Counter-Intelligence. 2013;**26**(3):453-481

[17] Wasielewski C, Sam Nivethan VJ. Securium Solutions – Snort Setup for IDS. Feb 2023. Available from: https://securiumsolutions.com/setting-up-snort-ids-basic-configuration/

[18] Chitrakar R, Chuanhe H. Anomaly detection using support vector machine classification with k-medoids clustering. In: 2012 Third Asian Himalayas International Conference on Internet. USA: IEEE; 2012. pp. 1-5. DOI: 10.1109/AHICI.2012.6408446

[19] Reif M, Goldstein M, Stahl A, Breuel TM. Anomaly detection by combining decision trees and parametric densities. In: 2008 19th International Conference on Pattern Recognition. USA: IEEE; 2008. pp. 1-4. DOI: 10.1109/ICPR.2008.4761796

Chapter 3

Applications of AI and IoT for Advancing Date Palm Cultivation in Saudi Arabia

Maged Mohammed, Nashi K. Alqahtani, Muhammad Munir and Mohamed A. Eltawil

Abstract

Date palm cultivation is an essential part of Saudi Arabia's economy. However, it faces several challenges: water scarcity, improper farm management, pests and diseases, inadequate farming practices, processing and marketing, and labor shortages. Artificial intelligence (AI) and the Internet of Things (IoT) can help enrich crop management, enable predictive analytics, increase efficiency, and promote sustainability in date palm cultivation. Recently, interest in this sector has begun by applying the latest precision engineering technologies integrated with AI and IoT techniques to address these challenges. This chapter aims to provide an overview of the applications of AI and IoT-based technologies, such as sensors, ML algorithms, and data analytics, and their potential benefits and challenges in supporting date palm cultivation in Saudi Arabia. Specifically, the applications of AI and IoT in smart precision irrigation, smart systems, cold storage management, pest infestation prediction, and date fruit quality optimization. In addition, the potential economic and environmental benefits of using AI and IoT in date palm cultivation in Saudi Arabia and the challenges that need to be addressed to realize these benefits fully. The chapter provides insight into the latest developments and future directions for AI and IoT in date palm cultivation, providing valuable information for researchers and policymakers.

Keywords: artificial intelligence, artificial neural networks, machine learning, prediction, estimation, precision agriculture, intelligent systems

1. Introduction

The agriculture sector worldwide has experienced significant changes recently, which have altered daily farm work activities. Due to the decrease in profit of certain farming businesses, the intensification of agricultural operations has followed, which has resulted in a diversification of farm activities. However, due to increasing industrial regulation, the agriculture sector's mechanization and automation have significantly influenced the qualitative transformation of farmers' daily work activities and those of the entire farming community globally. The farming industry has transformed because of new machinery, technological advancements, genetically modified seeds, new fertilizers, and organic and sustainable farming. Innovative technologies are becoming increasingly crucial in the agriculture industry as a tool for sustainable development. Agriculture automation is a valuable strategy that has enhanced product quality while reducing production costs and manual labor and improving environmental sustainability. Automation in farming equipment, irrigation and fertigation systems, climate control, pest and disease management, and soil fertility contribute significantly to agricultural productivity [1–6].

Artificial intelligence (AI) can help farmers improve crop yields, reduce biomass waste, and make better decisions. To build precise maps of crops and soil, AI is used to analyze data recorded from sensors, drones, and satellites. The data is used to optimize fertilizer, irrigation application, and pest control. AI-driven irrigation systems, for instance, regulate watering schedules based on weather patterns and soil moisture levels, lowering water usage and boosting agricultural yields. AI technology is also used for crop monitoring purposes. It examines satellite or drone-shot images of crops to spot problem regions. For instance, it can spot disease or nutritional deficiency symptoms before they appear to the naked eye. Farmers are then able to take measures before the issue gets worse. AI also analyzes historical data on weather patterns, soil conditions, and crop yields to predict future outcomes. This information can be used to optimize planting schedules, predict crop yields, and identify areas at risk of crop failure. Drones and robots with AI capabilities can perform many on-farm tasks, from planting seeds to harvesting crops. As a result, less manual labor is required, and crop productivity increases. The health and well-being of livestock can be monitored using AI technology. It can, for instance, examine sensor data from animals to find symptoms of certain diseases [7–15].

Internet of Things (IoT) is a network of devices that sense real-time conditions and then trigger actions to respond, but many IoT applications require more complex rules to link triggers and responses. IoT applications in agriculture use sensors and other connected devices to collect real-time data and automate farming operations, leading to better efficiency, decreased costs, and improved yields. The agriculture sector is undergoing a revolution because AI and IoT technologies provide farmers access to real-time data and insights that help them make precise decisions. Integrating AI and IoT in agriculture improves farm productivity, reduces waste, and optimizes crop yield. In order to collect real-time data on crop health, soil moisture, and meteorological conditions, IoT technology is being employed in agriculture. This data can be collected by sensors placed around a farm and transmitted to a central database for AI algorithm analysis. This enables farmers to decide on irrigation, fertilizer, and other farming practices. AI and IoT in agriculture help lessen the environmental effects of farming in addition to increasing crop yields and minimizing waste. Farmers can reduce the use of pesticides and fertilizers, consume less water, and produce less greenhouse gas emissions by optimizing their farming operations [5, 16–22].

The AI and IoT paradigms have seen widespread adoption by numerous businesses in recent years. Modern precision agriculture practices are largely mechanized and are integrated with effective and well-developed AI and IoT technologies. Although the concept of the IoT and AI is not new, it has recently gained massive popularity, mainly because of the upgradation in hardware technology over the past decade. AI and IoT technologies rapidly transform the agriculture sector through increased production, cost savings, and sustainability. Moreover, contemporary precision agriculture research has recently amalgamated with machine learning (ML) techniques to devise innovative solutions for agricultural challenges. Robots, drones, remote sensors, computer and satellite imagery, constantly evolving ML models, and analytical equipment are used in the AI and IoT to monitor crop and livestock, storage rooms, macro-, and microclimate, survey, and field mapping in real time, as well as to provide data to farmers for logical farm management strategies that will not only simple, time, and resources saving but also improve crop production. These technologies are, therefore, increasingly being employed in the agriculture sector to support more environment-friendly on-farm operations and improve ecological sustainability, all while maintaining the financial stability of farming enterprises [15, 23–27].

Date palm (*Phoenix dactylifera* L.) cultivation is an essential economic activity in many arid regions, including Saudi Arabia. Despite its economic importance, date palm cultivation faces several challenges, including water scarcity, pests and diseases, climate change, and the lack of postharvest processing technologies [28]. To address these challenges, there is a need for innovative and advanced solutions that can improve crop management and increase water use efficiency in date palm cultivation [29]. Recent AI and IoT advances can transform agriculture by providing real-time data on fruit quality, tissue culture systems, crop health, soil moisture content, and meteorological conditions [30]. In addition, using AI and IoT technologies in date palm cultivation can bring several environmental benefits, including optimized water use, reduced pesticide use, and improved soil health. Date palm cultivation is also water-intensive and can strain local water resources. AI and IoT technologies can bring several environmental benefits to date palm cultivation. One potential benefit of using AI and IoT in date palm cultivation is water use optimization.

By installing sensors in the soil and on the trees, farmers can monitor soil moisture levels and adjust irrigation accordingly. AI algorithms can analyze the data collected by these sensors and recommend when and how much water to apply. This can help reduce water waste and increase water use efficiency, particularly in areas with limited water resources. Another potential benefit is the reduction of pesticides and fungicides. IoT devices, such as drones equipped with cameras and sensors. Can detect pest and disease infestations early, allowing farmers to act before the infestation becomes widespread. AI algorithms can also analyze data from these devices to recommend the most effective and least harmful pesticides and fungicides. This can reduce the amount of these toxic chemicals used in date palm cultivation, positively affecting local ecosystems. Using AI and IoT in date palm cultivation can also improve soil health. By monitoring soil bionomics, moisture content, nutrient levels, and other factors, farmers can adjust their fertilization practices to ensure that they provide their trees with the nutrients they need while minimizing excess fertilizer application. This can help reduce soil degradation and improve overall health [31–34]. However, there is a lack of review research and book chapters on the application of AI and IoT in date palm cultivation, particularly in Saudi Arabia. Previous books and research on the application of AI and IoT in agriculture have focused primarily on different crops, with limited attention given to date palm cultivation.

Furthermore, most previous research has been conducted in temperate regions, where farmers face challenges that differ from those in arid regions, such as Saudi Arabia. Therefore, this chapter aims to address these gaps in knowledge by investigating the applications of AI and IoT-based technologies, such as sensors, machine learning algorithms, and data analytics, and their potential benefits and challenges for promoting sustainability and advancing date palm cultivation in Saudi Arabia. In addition, the chapter discusses the challenges that must be addressed to realize the full benefits of using AI and IoT in date palm cultivation. By providing insight into the modern developments and future trends for AI and IoT in date palm cultivation, the chapter provides valuable information for researchers and policymakers interested in using these technologies in arid regions.

The rest of the chapter is structured as follows: First, we introduce the importance and challenges of date palm cultivation in Section 2, Section 3 describes AI and IoT technologies, Section 4 provides an overview of applications of AI and IoT in agriculture, Section 5 details benefits and application of AI and IoT in date palm cultivation, Section 6 indicates the challenges of implementing ai and AI and IoT in date palm cultivation, Section 7 suggests future opportunities for AI and IoT in date palm cultivation, and Section 8 concludes the work.

2. Importance and challenges of date palm cultivation

Date palm is commonly grown in arid- and semiarid regions of the world on 1.31 million hectares and produces 9.82 million tons of fruit yearly. In Saudi Arabia, date palm is a major fruit crop in dry regions of North Africa, the Middle East, and parts of Asia, which provides food, nutrition, and building materials to the inhabitants and other byproducts. More than 120 million date palm trees are worldwide, and more than 84 million trees are grown in the Arab world (Egypt, Iraq, Saudi Arabia, Algeria, Morocco, Tunisia, and the United Arab Emirates). Arab countries have 70% of the world's date palm trees, contributing 67% of global production. More than 23 million date palm trees are grown in Saudi Arabia on 152,734 hectares of land, which yield 1.57 million tons of dates annually [35–38].

The cultivation of date palms in Saudi Arabia is significant for several reasons, including its economic importance. The date palm industry is labor-intensive and helps both males and females by generating revenue and jobs. Due to increased employment prospects in rural areas, widespread migration to cities is lessened. Women have a significant role, especially during the palm propagation stages (using *in vitro* or conventional methods) and postharvest stages, including packaging and marketing. Date production and trading help local economies and serve as a source of revenue for farmers and exporters. Over the past few decades, Saudi Arabia's agricultural sector, particularly the date palm sector, has experienced tremendous growth and support [39–43].

Many countries in the Middle East and North Africa, such as Saudi Arabia, Iran, Iraq, Egypt, and Tunisia, rely substantially on the export of dates. In 2021, Saudi Arabia was the leading global exporter of fresh or dried dates, with an export value of about 322.84 million USD [44]. However, the production and profitability of Saudi Arabia's date palm producers are constrained by several challenges. These challenges include water scarcity due to the depletion of groundwater, soil degradation by salinization, soil erosion, and desertification due to the loss of vegetation cover by overgrazing and overharvesting of wood for fuel, insect pest infestations, disease, an insufficient number of processing and packaging facilities and technologies, environmental pollution, and a decline in consumer demand for date fruit [29, 38, 45–47].

In Saudi Arabia, water scarcity is one of the biggest challenges that date palm producers face. The country's freshwater reservoirs are scarce, and most water sources are saline. As a result of the country's limited water resources, the agricultural sector consumes ca. 90% of the water. Due to the substantial water requirements of date palms, farmers have experienced decreased yields and elevated production costs. Extended droughts are common in the country, limiting the amount of water used for agriculture. In addition, because of urbanization and population growth, there is an increase in water consumption, creating competition for the limited supply of water resources. Farmers of date palms who rely on irrigation to sustain their crops are under strain because of this situation.

Additionally, date palm farmers lack effective irrigation systems, such as drip and subsurface irrigation systems. Many farmers employ ineffective traditional flood irrigation systems that cause significant water losses through evaporation and runoff. The farmers of date palms who use this method also contribute to soil salinization, worsening the irrigation water shortage. To address this problem, date palm farmers use desalinated seawater for irrigation, which is costly and unavailable to all farmers. The Saudi government has significantly invested in irrigation systems and technologies to encourage effective water use in agriculture to address this issue [48–50].

Desertification and soil erosion are further threats to the sustainability of date palm production. The process of topsoil being removed by wind or water and leaving unusable land that cannot support plant growth is called soil erosion. On the other hand, desertification refers to degrading formerly productive land into desert-like conditions because of natural or human-caused factors such as excessive grazing, deforestation, and unsustainable farming methods. In Saudi Arabia, several causes, such as climate change, excessive groundwater use, and unsustainable farming methods, have contributed to the deterioration of soil health and desertification. The country is especially prone to these issues because of its arid climate and constrained water supplies [51–57].

Another challenge date palm farmers face in Saudi Arabia is the prevalence of pests and diseases. Date palms are susceptible to various pests, including red palm weevils, long horn borer, dust mites, fruit flies, scale insects, and fungal diseases. The red palm weevil is an invasive species that infects young and mature palm trees, damaging their vascular system. The weevil larvae bore into the tree trunk, feeding on the soft tissues and creating tunnels that damage the tree structure. The leaves of infested trees display wilting and yellowing symptoms, resulting in death [58–66].

Postharvest losses are estimated to be around 10% in Saudi Arabia, where date palm farming is an important industry. Limited postharvest facilities in the country are one of the key reasons for these losses. Postharvest facilities are essential for the proper storage, processing, and packaging of dates. They help prevent spoilage, minimize losses, and maintain the fruit's quality. However, due to various issues such as high costs, lack of knowledge, and inadequate infrastructure, many date palm farmers in Saudi Arabia do not have access to these facilities [67–71].

Another issue Saudi Arabian date palm growers deal with is a skilled manpower shortage. Many date palm farmers rely on skilled migrant labor during harvest and perform different cultural practices such as pollination, pruning, fertilization, irrigation, chemical application, etc. However, changes in immigration policies and geopolitical strains have made it difficult for farmers to access skilled labor when needed. This has resulted in labor shortages during critical periods, which can lead to reduced yields and financial losses for farmers. As a result, skilled labor shortages during critical periods reduced production and financial losses for farmers [72, 73].

3. AI and IoT technologies

AI refers to a group of technologies that allow computers to do a wide range of complex tasks, such as seeing, comprehending and translating spoken and written language, analyzing data, making suggestions, and more. It revolutionized agriculture by making farming more efficient, sustainable, and profitable. In addition, many businesses are embracing IoT, which offers simple means to collect and evaluate technical data to identify and improve the performance of numerous daily operations. The technological revolution reveals new challenges and issues with our current IoT technologies. New technologies, such as AI, 5G, and blockchain, promise to solve these challenges. We can create intelligent machines with the help of IoT and AI integration. These innovative automation technologies not only make monotonous tasks more accessible but they also make smart decisions without human assistance. The IoT and AI are two of the most significant technologies in the computing industry, completely transforming how we communicate with machines and our environment. It is speculated that ca. 64 billion AI and IoT devices will be available by 2025. AI and IoT technologies work effectively together and are at the top of the latest innovations influencing the information technology sector. The industrial and agribusiness sectors have benefited from the duet's redesign of traditional solutions [74, 75].

Machine learning (ML) is a subset of AI that involves training algorithms to make predictions or decisions based on training data. ML is a rapidly growing field with numerous applications and opportunities for scientific innovation. It is a branch of AI that involves the development of algorithms and statistical models that allow computer systems to automatically improve their performance on a specific task as they gain experience. ML aims to develop intelligent systems capable of learning from data and making predictions or judgments without being specifically programmed to do so. In IoT, ML is a technique that can be applied to analyze sensor data and forecast future events. Deep learning has become a potent ML method in recent years. Deep learning includes putting multiple-layer neural networks through training to learn how to represent data hierarchically. As a result, developments have been made in fields such as speech recognition, natural language processing, and image recognition. Numerous industries, including agriculture, healthcare, finance, marketing, and transportation, use ML in various ways. ML in agriculture has the potential to be applied in a variety of ways with exceptional results from weed, pests and disease detection, crop yield, and quality prediction, to data collection, providing insights, and livestock production forecasting. It can be used in the healthcare industry to find novel medicines and disease diagnoses. It can be applied to finance to assess risk and detect fraud. It can be used in marketing for personalized advertisements and customer segmentation. It can be applied to autonomous driving and traffic forecasting in the transportation sector [29, 76–81].

Natural language processing (NLP) is a field of AI that applies to the interaction between computers and humans using natural language. NLP is vital to contemporary AI systems because it enables machines to understand, interpret, and generate human language. The development of smart homes, smart cities, and other intelligent systems now has more opportunities since integrating NLP in IoT devices. Users can communicate more effectively and easily with IoT devices by adding voice recognition and natural language understanding capabilities. One of the most significant applications of NLP in IoT is voice assistants such as Amazon's Alexa, Google Assistant, and Apple's Siri to turn on/off lights, play music, set alarms, or order groceries online. Weather data analysis is an example of how NLP is used in agriculture. Farmers can decide on crop planting and harvesting dates, irrigation requirements, and controlling pests and diseases by evaluating weather patterns and forecasts. Farmers may track crop growth and development rates, as well as soil moisture levels with the aid of NLP algorithms. In order to help the farmers choose the best crops to cultivate and the best times to sell, NLP models can also be used to assess market trends and forecast demand. Farmers may find new customers and bargain for higher prices with the aid of NLP algorithms. Improved consumer-farmer communication is another area, where NLP can be employed. Farmers can learn more about consumer preferences by examining customer feedback analysis. Using NLP algorithms, farmers can communicate more effectively with each other, sharing their best agriculture practices and working together on research initiatives [82–86].

Computer vision is a field of AI that enables computers and machine systems to extract useful information from visual data (digital images and videos) and then take actions or make recommendations. This technology is widely used in IoT to enable devices to see and interpret the physical entity through object, facial, and gesture recognition, thus making them more intelligent and responsive. With the help of deep learning algorithms, computer vision systems can identify objects in real time and classify them into different categories. This capability has been used in various IoT applications, such as autonomous vehicles, security cameras, and smart home devices. Similarly, computer vision systems can extract valuable environmental information by analyzing images captured by cameras or sensors. For example, they can detect anomalies or changes in a scene, monitor traffic flow, or track the movement of people or objects. IoT has also used this technology for gesture recognition and human-computer interaction. Computer vision systems can enable users to interact with devices more naturally and intuitively by analyzing hand movements and gestures. This capability has been used in various applications, such as gaming consoles, virtual reality systems, and smart home devices. Many agricultural activities are being automated and optimized using computer vision technology. Crop monitoring, yield prediction, and insect and disease detection are a few applications of computer vision technology in agriculture [87–91].

Robotics technology is a branch of engineering that entails creating and programming robots and is crucial to advancing AI and the IoT. It is used in IoT to create intelligent machines that can exchange data and connect. Robotics is using machines to carry out jobs that are either harmful or too difficult for humans. How we interact with machines has changed dramatically because of the inclusion of robotics technology in AI and IoT, improving their efficiency, dependability, and autonomy. These robots' 24/7, nonstop operation, which is also utilized for data entry and processing tasks, minimize errors, and boost efficiency. Automation of robotic processes has been widely used in several sectors, including agriculture, manufacturing, healthcare, logistics, and finance. The development of autonomous robots is another way that robotics technology is used in AI. Robots operating independently of humans may use sensors, cameras, and other cutting-edge technologies to navigate their environment. Robotic technology is being employed more frequently in agriculture to improve the productivity and efficiency of farming operations. Many benefits are reported from using robots in agriculture, such as improved crop yields, lower labor costs, and increased accuracy [15, 92–95].

Edge computing represents a decentralized computing model that moves computation and data storage nearer to the requirement point, enhancing response speed and minimizing bandwidth consumption. In IoT, edge computing can process sensor data in real time at the network's edge. ML models used in AI are substantially trained using massive volumes of data. This training can be carried out locally *via* edge computing, reducing the need to send large amounts of data to centralized servers. By storing sensitive data on local devices rather than transferring it over the internet, edge computing can decrease network congestion, enhance privacy and security, and increase reliability. Edge computing can also be utilized to run AI models in real time, enabling rapid response and decision-making. Edge computing, for instance, can be utilized to analyze sensor data from a localized area in real time and identify possible problems before they escalate [96–101].

Cloud computing technology plays a crucial role in developing and deploying AI and IoT applications. It provides an infrastructure that allows organizations to store, manage, and process large amounts of data generated by IoT devices and AI applications. Within the cloud computing framework, the vendor hosts and provides infrastructure, data, and software as a service to the user. Scalability is one of the most significant benefits of cloud computing for AI and IoT. Because cloud-based services are easily scaled up or down according to demand, businesses and organizations can manage the massive amounts of data and processing power needed for AI and IoT applications.

Additionally, cloud computing offers an affordable solution for users who need to store and process large amounts of data. By utilizing cloud-based ML platforms, cloud computing also enables the development of AI and IoT applications. These platforms give programmers access to robust machine-learning algorithms, resources, and libraries that they can utilize to develop sophisticated applications.

Additionally, developers can scale up or down their ML models in response to demand using cloud-based ML platforms. Additionally, cloud computing offers a safe environment for processing and storing private data produced by IoT and AI applications. To safeguard customer data from cyber threats, cloud service providers have implemented several security measures such as firewalls, access controls, and encryption [102–106].

Blockchain technology is a decentralized and distributed ledger system that securely and transparently records transactions across multiple computers. It enables secure transactions between parties without the need for intermediaries. It can be used for secure device authentication, data sharing, and supply chain management. This technology can enhance AI and IoT systems' security, privacy, and interoperability. Increased security is one of the main benefits of adopting blockchain in AI and IoT. It secures data and transactions using cryptographic algorithms, making it difficult for hackers to tamper with or steal data. Blockchain can also be used to develop secure digital identities for users and devices, which can help restrict unauthorized access. Improved privacy is a benefit of using blockchain in AI and IoT. Users can choose who can access their data and maintain control over it. This technology is essential when sensitive data needs to be protected, such as in healthcare and financial applications.

Additionally, blockchain can improve the interoperability of various IoT and AI systems. It enables different methods to communicate with each other more easily by establishing a shared decentralized ledger of transactions. This can make it simpler for developers to create new applications and lessen the difficulty of integrating various systems [107–111].

4. Applications of AI and IoT in agriculture

By 2025, it is anticipated that global spending on intelligent, interconnected agricultural technology and systems, including AI and IoT, will triple in size, reaching $15.3 billion. Understanding how factors such as sunlight, weather, animal, bird, and insect movements, crop use of specific fertilizers and pesticides, and planting and irrigation cycles affect crop production is a perfect subject for machine learning. Excellent data has never been vital for determining a crop cycle's profitability. For this reason, farmers and the agricultural sector are stepping up their data-centric strategies and broadening the scope and scale of the application of AI and IoT technology to improve crop yields and quality [112–116].

Crop yields are being increasingly optimized using AI and IoT. Farmers can understand more about the health of their crops and make intelligent decisions about water and fertilizer requirements and disease and pest control by utilizing data from sensors and other connected devices. Precision farming is a significant area, where AI and IoT are used in agriculture. This entails creating precise field maps and real-time crop growth monitoring using data from sensors, drones, and other sources. Farmers can identify areas that need water, fertilizer, or disease/pest control by examining this data and areas that are vulnerable to disease or pest infestation. Due to their increased ability to deploy resources effectively and efficiently, crop yields increase and costs decrease [117–120].

Predictive analytics is yet another approach AI and IoT use in agriculture. Farmers can use machine learning algorithms to accurately predict future yields by analyzing historical data on weather patterns, soil conditions, and crop performance. This enables them to plan for planting and harvesting crops and make smart decisions about pricing and marketing. Before a vegetation cycle even begins, it is now possible to know the potential yield rates of a field using AI and IoT technology. The potential crop yield can be estimated using machine learning techniques to analyze 3D mapping, sensors' soil analysis data, and drone-based soil color data. Farmers can automate the irrigation schedule of their crops based on real-time data on soil moisture content *via* remotely connected devices, such as smart sprinklers and soil sensors. Satellite-based thermal-infrared imaging remote sensing AI and IoT technology also monitor irrigation rates and crop water requirements. Automating tasks, such as fertilization is another possible application of AI and IoT in crop farming. Farmers can apply fertilizer more precisely based on real-time data on soil nutrient levels *via* GPS-connected fertilizer spreaders. The technology is used to apply fertilizer variably on most needy soil areas.

Similarly, the normalized difference vegetation index images recorded through drones or satellites can be used to apply variable nitrogen application at different crop growth stages. Variable seed rates of different crops can be estimated by scanning the electric conductivity of the field. The seed spreader is then connected to the GPS-kit with the field electric conductivity map, guiding the spreader to apply seed variably. Large-scale agricultural firms turn to robotics when they cannot find enough skilled workers. Self-propelled robotics machinery that can be programmed to apply seed and fertilizer evenly along each row of crops lowers operational costs and increases crop yield. Farmers can detect symptoms of disease or stress before they are noticeable to the naked eye, for instance, by employing computer vision algorithms to examine photographs of plants. This enables them to take corrective action before the problem worsens, resulting in healthier plants and better-quality produce. Farmers employing AI and IoT technologies can predict and detect disease/pest infestations before they occur by combining drone infrared camera data with sensors on fields that can monitor plants' relative health.

Moreover, the AI and IoT systems can identify disease/pest-affected areas by combining intelligent sensor data with visual data streams from drones. Farmers can gain more knowledge about the health of their crops and decide how to allocate resources and control pests/diseases by analyzing data from sensors and other connected devices. Crop quality can also be improved as well with the use of AI and IoT [121–131].

All agricultural supply chains have adopted track and traceability, and this trend is expected to continue. A well-managed track-and-trace system increases visibility and control throughout supply chains, which reduces inventory shrinkage. Modern AI and IoT-based track-and-trace systems can distinguish between batch, lot, and container-level material assignments in inbound shipments. Most cutting-edge track-and-trace systems rely on sophisticated sensors to record data for each shipment. Agricultural supply chains and shipments are increasingly using AI and IoT sensors. AI and IoT systems show different marketing scenarios to farmers to get the maximum return on their produce. When deciding on pricing strategies for a particular crop, price forecasting for crops based on yield rates that help forecast total volumes produced is crucial. Understanding yield rates and quality standards enables agricultural businesses to negotiate effectively for the best harvest price. The pricing strategy is determined by analyzing the total demand for a crop to determine whether the price elasticity curve is inelastic, unitary, or highly elastic [7, 118, 132–135].

One of the fastest-growing applications of AI and IoT in agriculture is monitoring livestock health, including daily activity and food intake. Various aspects of livestock management and monitoring, such as behavior, detection, counting, identification, grazing tracking, health issues, estimating the herd distribution, etc., can be achieved using AI and IoT technologies. The best way to care for livestock over the long term is to understand how each livestock responds to diet and boarding conditions. Producing more milk requires AI and IoT to comprehend what keeps cows happy and contented daily. AI and IoT technologies reduce the chances that domestic and wild animals may accidentally destroy crops or commit a break-in or burglary at a remote farm. Farmers can secure the perimeters of their fields and buildings through image analysis powered by AI and machine learning algorithms [136–141].

5. Benefits and application of AI and IoT in date palm cultivation

Advances in technology have led to the integration of AI and IoT into the date palm farming sector, aiming to improve yield, reduce costs, and increase efficiency. One recent development in AI and IoT in date palm cultivation is using sensors to monitor soil moisture. These sensors are connected to a central system that uses AI algorithms to analyze the soil moisture data and determine the time and amount of irrigation water requirement. This approach has been shown to reduce water consumption by up to 30–60% and improve date palm yields. Another application of AI and IoT in date palm cultivation is using drones for crop monitoring. Drones equipped with cameras and other sensors collect data regarding plant health, growth rates, and other factors affecting palm growth and yield. This information can then be analyzed using ML algorithms to identify patterns and predict future crop performance. In addition to these applications, AI and IoT are also used to optimize fertilizer usage, predict weather patterns, and automate the harvesting time of different date palm varieties. For example, autonomous robots with AI algorithms can harvest dates more efficiently than human laborers, reducing costs and increasing productivity [20, 29, 63, 65].

In many parts of the world, date palm cultivation is a significant economic activity. Using AI and IoT technology in date palm farming can have several economic benefits. Enhanced crop management efficiency is one potential benefit. AI-powered sensors can monitor soil moisture, aerial temperature, humidity, and other environmental variables that impact the development and production of date palms. Date palm growers can optimize irrigation schedules, fertilizer applications, and disease and pest management strategies by analyzing AI and IoT data, leading to higher crop yields and lower input costs. Improved quality control is another potential benefit. AI algorithms can examine images of date palm trees to identify signs of water stress, diseases, or nutrient deficiencies. This can help farmers identify problems early and resolve them before they worsen. IoT devices can also track the shipment of date palm offshoots from the field to the market, ensuring that they are treated carefully and adhere to quality standards. Increased market access is a third potential advantage. Farmers can produce higher-quality dates that fetch higher prices in domestic and foreign markets by applying AI and IoT technology to enhance crop management and quality control.

Furthermore, IoT devices can deliver real-time information about market demand, allowing farmers to modify their production strategies accordingly. AI techniques made identifying different date palm varieties possible through leaf and fruit image scanning. In general, date palm farming could benefit from AI and IoT technology by increasing productivity, enhancing quality assurance, and giving farmers more market access. The economic benefits of these technologies are expected to significantly increase as they develop and become more accessible [19, 20, 30, 32, 38, 43, 142–144].

The following are some of the applications of AI and IoT that have been employed for advancing date palm cultivation technology:

5.1 Date palm irrigation management

Fresh water is an urgent priority in semiarid- and arid regions. With the steady increase in population, water is urgently needed to irrigate palm trees and increase the production of dates, food products rich in nutrients necessary for human health. The reclaimed and desalinated water can be used for irrigation, but these technologies' high energy and cost hinder this irrigation utilization [19, 145]. There is a need for intelligent irrigation and standalone photovoltaic systems, and new smart irrigation techniques to ensure sustainable energy and water for agriculture [20, 21, 146]. Mohammed et al. [29] implemented AI to predict optimum water and energy requirements for solar-powered sensor-based microirrigation systems. This study is a good example of how AI can improve agricultural practices. The study also found that the optimum water use efficiency was achieved when the maximum setpoints of irrigation control were adjusted at the field capacity and by adjusting the minimum setpoints at 40 of the available water for the subsurface irrigation system. The optimum yield was achieved by adjusting the minimum setpoints for subsurface irrigation, subsurface drip irrigation, and bubbler irrigation, respectively. Several ML algorithms were used in the study, including support vector regression (SVR), long short-term memory (LSTM) neural network, linear regression (LR), and extreme gradient boosting (XGBoost) were developed and validated for predicting the optimum irrigation water and solar energy requirements based on the limited meteorological data (average temperature, RH, wind speed, and solar irradiance) and date palm age for each microirrigation system used. The study evaluated the performance of the ML models using three performance metrics of root mean square error (RMSE), coefficient of determination (R^2), and mean absolute error (MAE). The dataset was prepared at various levels for 4 years to train and test the prediction models, and the fifth year was used to validate the performance of the most suitable model. The evaluation of the ML models indicated that the LSTM and XGBoost models were more accurate than the SVR and LR models in predicting the optimum irrigation water and energy requirements. The validation results showed that the LSTM model could predict the water and energy requirements for all irrigation systems with R^2 values ranging from 0.90 to 0.92 based on date palm age and limited meteorological variables. The authors stated that the benefits of implementing AI in sustainable farming include predicting optimum water and energy requirements for sensor-based microirrigation systems powered by solar PV, contributing to sustainable farming practices. They also highlighted the potential significance of AI in effectively overseeing irrigation water scheduling. AI could achieve this by handling gathered data and comprehending the evolving weather patterns, soil, and plant conditions throughout the cultivation phases. The LSTM model they created could serve as a potent instrument for supervising water allocation in date palm cultivation [29].

Consequently, AI's influence extends to water conservation in irrigation, fostering plant development, and augmenting crop yield. In addition, the study's results have significant implications for sustainable agriculture in arid regions. By using AI to predict the optimum irrigation requirements, farmers can save water and energy while still achieving optimal yields. This is essential for ensuring the long-term sustainability of agricultural production in arid regions [29].

A previous study highlighted the benefits of using the Internet of Things (IoT) in agriculture, especially irrigation management. Mohammed et al. [38] developed an automated system for scheduling irrigation using a cloud-based IoT platform, which positively impacted the yield of date palm and water use efficiency using FTTT (If This Then That) interface and Ubidots platform. **Figure 1** shows the main components of the cloud-based IoT platform used to monitor the meteorological variables in real time and control the irrigation water scheduling of the irrigation systems. The figure illustrates the flow of data from the sensors to the cloud platform and the control of the irrigation system through the Arduino UNO board and IFTTT interface. The meteorological variables were collected by the sensors and transmitted to the Arduino microcontroller in real time through the Wi-Fi module. The microcontroller

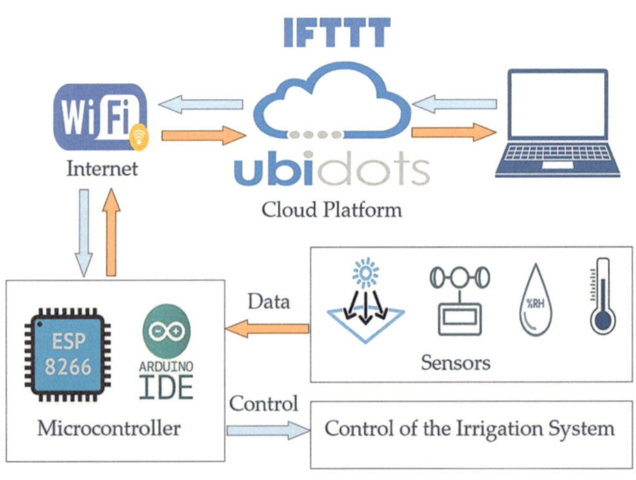

Figure 1.
A simple diagram for the components of the cloud-based IoT system employed for real-time monitoring of the meteorological parameters of the study area and managing water scheduling for date palms.

then sent the data to the Ubidots platform, where the user could access the real-time data through the graphical user interface using the private channel. The electronic relays of contactors, electronic valves, and irrigation pumps were controlled using the IFTTT interface to schedule the irrigation water of the irrigation system.

Implementing IoT procedures and creating sophisticated sensors for smart agriculture exert a significant influence on both crop yield enhancement and the preservation of irrigation water. Furthermore, integrating cloud computing and IoT advancements has bolstered the interaction and remote oversight between users and their agricultural operations. This enhancement permits multilayered cloud IoT and computing frameworks to orchestrate, supervise, and administer crop cultivation within a comprehensive automated structure. This has the potential to tackle the challenges posed by limited water availability and insufficient labor in the agricultural sector. The operation was overseen through a cloud-based IoT platform, allowing users to remotely observe the farm and retrieve pertinent meteorological information. This data empowered users to make informed decisions, considering the irrigation microcontroller's existing parameters. The IFTTT interface seamlessly integrated with the irrigation hardware, introducing functionalities that managed irrigation valves and pumps or dispatched SMS notifications to users based on their predefined actions. The IoT system optimized water use in date palm cultivation and improved yield and water use efficiency. In addition, the Ubidots platform was used in this study to monitor the meteorological variables data. The platform allows users to connect, visualize, and analyze data from various sources, including sensors, devices, and applications. The Ubidots platform provides a graphical user interface (GUI) that allows users to create custom dashboards, charts, and widgets to visualize data in real time. The platform also provides tools for data analysis, including statistical analysis, machine learning, and predictive analytics. This study used the Ubidots platform to collect and store real-time meteorology measurements on the farm to analyze and visualize the irrigation parameters [38].

Mohammed et al. [20] employed cloud-based IoT solutions to control a modern subsurface irrigation system in date palm farms in the arid region of Saudi Arabia, which improved irrigation management. They designed and constructed a fully automated controlled subsurface irrigation system (CSIS) and validated its performance to monitor the irrigation process remotely. An efficient control system for subsurface irrigation utilizing marvelous cloud computing and IoT capabilities was used to manage date palm water. The user can be automatically notified by either a short or email message. The optimum water per tree can be applied by controlling the subsurface irrigation system in a date palm field. The methodology used in this study involves designing and implementing a cloud-based CSIS for date palm trees. CSIS is an IoT-based system that employs cloud computing and various sensors to monitor and control the subsurface irrigation system for date palm trees. A sensor-based subsurface irrigation scheduling (S-BIS) was considered. Based on the data received from the sensors, the amount of water can be scheduled. The measured data from sensors is uploaded to the ThingSpeak cloud platform for analyzing and sending the decision to the subsurface irrigation system. **Figure 2** shows the designed system that used the direct measurement of volumetric water content to make irrigation decisions, meanwhile monitoring different factors such as ambient air temperature, relative humidity, solar intensity, wind speed, and water flow rate per minute. The results indicated that the automatically irrigating date palm trees controlled by S-BIS were more efficient than the time-based irrigation scheduling (T-BIS). The amount of irrigation water was reduced by 64.1% and 61.2% based on S-BIS and T-BIS, respectively, compared to traditional surface irrigation

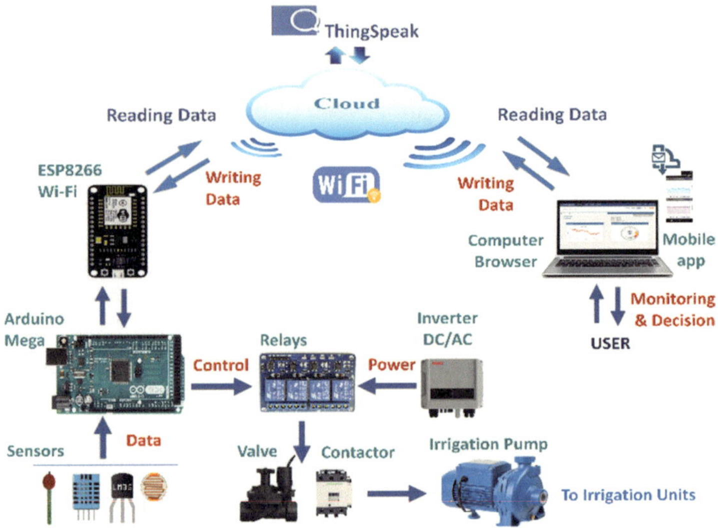

Figure 2.
A schematic diagram for the IoT-based control system for a smart subsurface irrigation for enhancing irrigation management of date palm.

(TSI). The yearly sum of irrigation water employed for CSIS utilizing the S-BIS technique, CSIS using the T-BIS approach, and TSI stood at 21.04, 22.76, and 58.71 m^3/tree, respectively. When integrated with the S-BIS methodology, the devised CSIS approach yields favorable outcomes regarding irrigation water administration and a subsequent improvement in date palm fruit yield within arid regions [20].

The IoT-based control system efficiently schedules the irrigation water to administer to the date palm at different intervals, relying on data from various sensors. The system collects measurements from these sensors, transmits the data to the ThingSpeak cloud platform for analysis, processes the information in the cloud, reaches conclusions, and then implements these conclusions into the subsurface irrigation system. The benefits of using IoT technology in this study include more efficient water management, improved crop yield, and reduced water waste. Additionally, the system can be remotely monitored and controlled, reducing the need for manual labor and increasing the accuracy of date palm irrigation management [20].

5.2 Tissue culture systems management

The smart *ex vitro* acclimatization systems (SEVAS) for tissue culture plantlets aim to minimize the initial shock of newly regenerated *in vitro* plantlets. This benefit decreases their mortality and improves their growth characteristics. In addition, the potential advantages of using SEVAS for tissue culture plantlets in agriculture include reduced production costs, reduced manual labor, enhanced product quality, and improved environmental sustainability. Utilizing automation is a pragmatic approach, particularly considering the extensive and time-consuming nature of *in vitro* propagation. This is especially true when dealing with limited outputs during the acclimation phase due to the mortality of plantlets. The benefits of automating the acclimatization process also encompass lower contamination risks and reduced labor expenses. Contemporary precision agriculture methods, such as glasshouse technology, are predominantly characterized by automation. In contrast, the combination of information technology and IoT solutions has advanced significantly, ensuring effectiveness and efficiency [43, 147, 148].

Mohammed et al. [43] designed an IoT-based automated system for the SEVAS to acclimate tissue culture plantlets. The designed system uses IoT technology to monitor and control the environmental conditions of the glasshouse, including temperature, humidity, and light intensity. The system also includes a feedback mechanism that adjusts the environmental conditions based on the real-time data collected by the sensors. The advantages of employing IoT technology in this research encompass the immediate tracking of crops and microclimate, surveying and mapping fields, and providing data to farmers for implementing well-founded strategies in farm management. These strategies aim to enhance efficiency, conserve time and resources, and elevate crop yield. Furthermore, IoT technology is progressively gaining traction within agriculture, facilitating the adoption of environmentally friendly on-farm methods and fostering improved ecological sustainability.

Figure 3 provides an overview of the IoT-based control and monitoring system for the SEVAS in the study. The figure shows the primary components of the system, comprising sensors, a microcontroller, an internet connection, the IoT platform hosted on the cloud, control apparatus, and web-based applications. These components are interconnected and harmonized to facilitate control and monitoring functionalities. The figure also depicts the trajectory of data from the sensors to the cloud-based IoT platform and the control devices responsible for overseeing the microclimate variables within the E-VAS system. The cloud-based IoT platform stores and analyzes the data and sends control signals to the control devices, which include relays for controlling the heating unit, cooling units, ultrasonic humidifier, and water pump and irrigation valves. The control devices adjust the SEVAS microclimate factors based on the sensors' data and the control signals from the cloud-based IoT platform [43].

5.3 Cold storage management

Cold storage is essential in food, vegetables, and fruit preservation. Refrigeration in remote areas away from the electricity grid needs an off-grid power system.

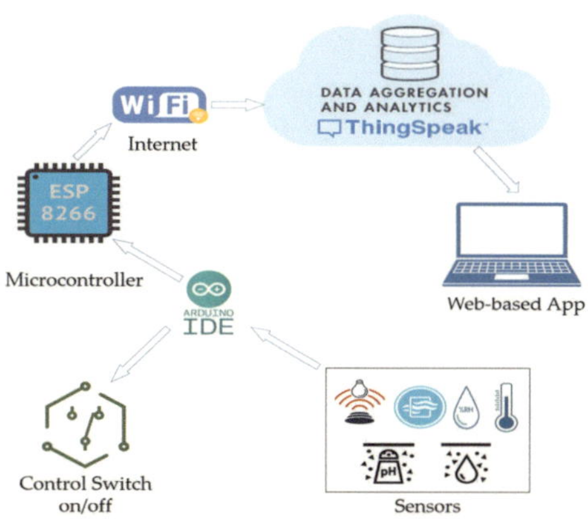

Figure 3.
A simple diagram for the components of the IoT-based controlling and monitoring system of the SEVAS for tissue culture plantlets.

Photovoltaic (PV) solar energy is an important power source for operating off-grid refrigeration. Due to a reduction in PV system cost, solar-powered refrigerators have become more economical [30, 71, 149, 150]. Refrigerators are considered one of the types of equipment that consumes a significant amount of electricity. Hence, reducing energy consumption and efficient systems are most important to reduce greenhouse gas emissions and the costs of PV systems [30, 151]. Eltawil et al. [152] developed and evaluated a machine learning-based intelligent control system (ICS) using artificial neural networks (ANN) for the performance optimization of solar-powered display refrigerators (SPDRs). The SPDR functioned initially at a consistent frequency of 60 Hz, and subsequently, it was operated at various frequencies ranging from 40 to 60 Hz. An integrated ANN-based ICS facilitated this frequency adjustment with a variable speed drive. An independent PV system provides the energy necessary for its operation. The performance of the newly developed SPDR was assessed and contrasted against its performance under a conventional control system (TCS). These evaluations were conducted at refrigeration temperatures of 1, 3, and 5°C, which align with ambient temperatures. The researchers employed an ANN-based regression model to enhance the SPDRs. The ANNHUB software's ANN technique created an optimal predictive model. This model was used to forecast the requisite power and ideal frequency for the SPDR, leveraging training data.

The Levenberg–Marquardt algorithm was employed in the training phase, allocating 75% of the data for training and 25% for testing. This algorithm optimized the weights in a composite estimation of outputs, thereby refining the prediction model. **Figure 4** shows the ANN architecture used in this study had a three-layered network consisting of one input layer, a hidden layer, and an output layer. The input layer has six nodes of six independent variables: target temperature (T1), ambient temperature (T2), cabinet temperature (T3), solar radiation intensity, and temperature differences of T2-T1 and T2-T1. The optimum hidden layer was one and contained eight nodes. The output layer had a single dependent variable, frequency, which was also standardized. The outputs were the optimum frequency to control the compressor speed and the required power. The approach used provided a more efficient and reliable means of control. The study suggests that using ML to optimize the performance of solar-powered refrigerators can lead to several potential benefits. For example, it can help

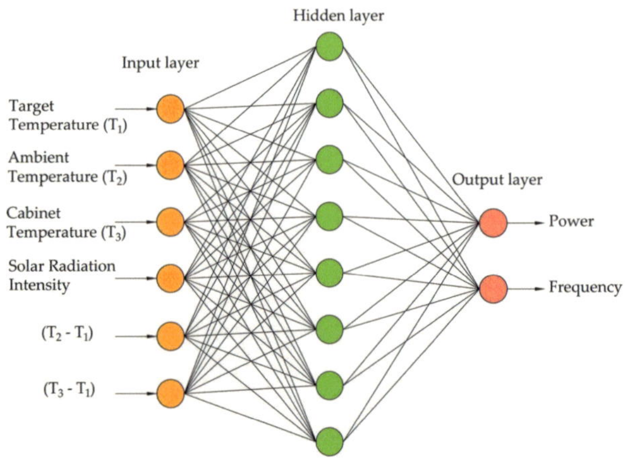

Figure 4.
The ANN architecture of the intelligent control system to optimize the performance of solar-powered refrigerators.

improve the reliability and efficiency of the control mechanism, save the required solar PV energy, and provide a basis for designing and optimizing PV-powered refrigeration systems. Furthermore, the research demonstrated that this approach can be extended to other refrigeration systems, offering a more effective and dependable regulation method [152].

IoT technology for managing cold storage facilities provides real-time system environment monitoring, allowing timely responses to any issues. This technology can give reliable data on the quality of food products during their storage duration, which can help with intelligent food quality management. Furthermore, the application of IoT technology in cold storage revolves around monitoring factors influencing the quality of stored products, with the goal of safeguarding them against potential contamination arising from external conditions. Among these crucial factors, cold storage rooms and warehouses focus on tracking parameters such as relative humidity (RH), temperature, alcohol gases, and light [71, 153].

Mohammed et al. [30] designed a smart IoT-based control system to manage cold storage facilities remotely. This study is a good example of how IoT can improve food safety and quality control. The study found that the IoT-based control system could precisely control the modified cold storage room, provide reliable data about the interior microclimate atmosphere, and send the necessary alerts in an emergency. This indicates that the IoT-based control system can improve the safety and quality of food products stored in cold storage rooms. The study also found that the IoT-based control system had no significant effect on the quality of date fruits stored in the modified cold storage room compared with the traditional cold storage room. This indicates that the IoT-based control system can maintain the quality of food products stored in cold storage rooms without affecting their taste or nutritional value. The study's results have significant implications for the food industry. By using IoT to monitor and control cold storage rooms, the food industry can reduce the risk of food spoilage and improve the quality of their products [30, 71]. **Figure 5** shows the main components of the designed IoT-BC. The figure illustrates the various elements of the system, including the IoT microcontroller, sensors, and cloud platform. The IoT microcontroller collects data from the sensors, which monitor various parameters such as temperature, humidity, and light. The collected data is then transmitted to the cloud platform, where it is analyzed in real time. The cloud platform is also responsible for sending notifications to the user in case of any issues that may arise [30].

5.4 Postharvest management

Climate change positively impacts date palm fruit growth and development by delaying fruit ripening, reducing color development and quality, inadequate pollination, fruit sunburn, poor fruit quality and fruit set, and lowering fruit yield [154, 155]. Date palm fruits can be ripened artificially using controlled temperature and relative humidity, such as the fruit drying process. Mohammed and Alqahtani [144] designed an automated sensor-based artificial ripening system (S-BARS) integrated with ultrasound pretreatment for unripe Khalas Biser fruits of date palm. They developed a straightforward technique for data acquisition and system control. Data on temperature and relative humidity were monitored using six DHT22 sensors. An Arduino Mega board collects the data sent by these sensors. **Figure 4** shows real-time data acquisition for temperature and relative humidity inside the treatment chamber of the designed S-BARS integrated with Arduino and Excel.

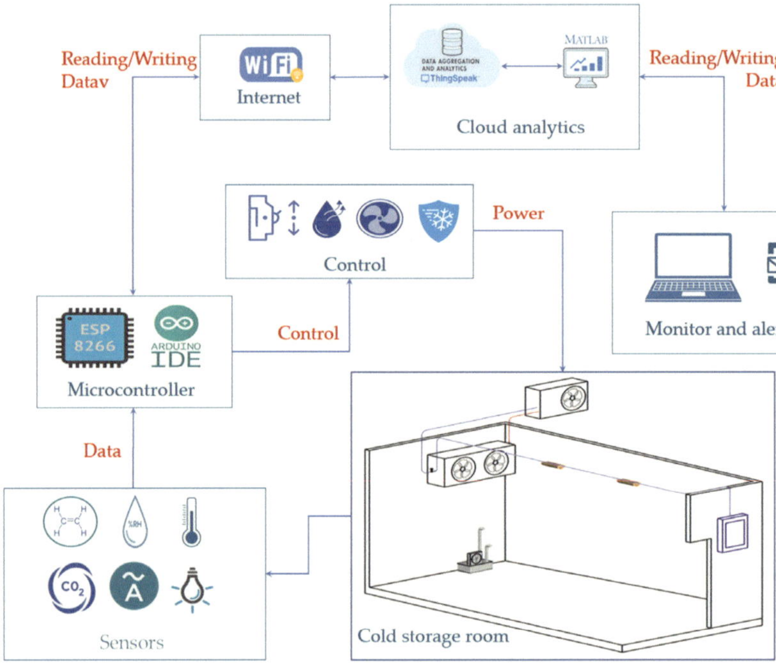

Figure 5.
A schematic diagram of IoT-based control system for cold storage room.

The control system in the study is an automated sensor-based artificial ripening system (S-BARS) that combines ultrasound pretreatment with automated sensors to enhance the ripening process of date fruits. The S-BARS are controlled by an open-source microcontroller board (Arduino Mega) and three relays (RL1, RL2, and RL3) that control the heating unit, ultrasonic humidifier, and main power of the S-BARS. The system also includes six DHT22 sensors that collect data on temperature and relative humidity (RH) and send the data to the Arduino Mega board's open-source microprocessor (ATmega328P). The acquired data is displayed in real time on a liquid crystal display (LCD) and stored in Microsoft Excel using the PLX-DAQ Excel Macro. The control system allows for precise monitoring and control of the ripening process, which can improve the quality and yield of date fruits [144].

Mohammed et al. developed ANNs-based models for predicting date fruit quality attributes based on their electrical properties during cold storage. This study is a good example of how machine learning can be used to improve food safety and quality control. The study found that ANNs were more accurate than multilinear regression (MLR) models in predicting the physicochemical properties of date fruits during cold storage. Therefore, ANNs can be used to develop nondestructive methods for predicting the quality of date fruits. Ensuring a consistent provision of premium fruits to meet market requirements is paramount. The research also identified that the most effective prediction model utilizing ANNs comprised an input layer with 14 neurons, a single hidden layer with 15 neurons, and an output layer with four neurons. The study's findings have significant implications for the food industry. By using ANNs to predict the quality of date fruits, the food industry can reduce the amount of food waste, improve the quality of their products, and meet the demands

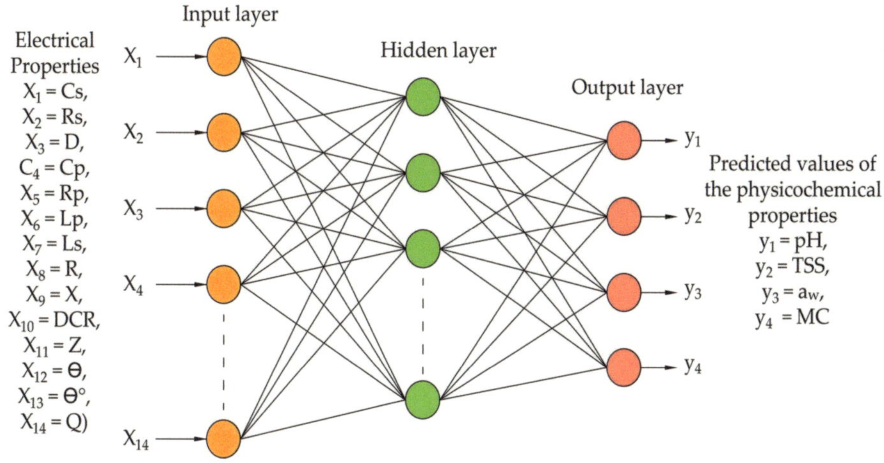

Figure 6.
A simple block diagram of the employed ANNs prediction model for prediction of fruit quality based on their electrical properties during cold storage.

of consumers. The study is a valuable contribution to food safety and quality control. The study's results have significant implications for the food industry, and food manufacturers and policymakers should consider them [78]. **Figure 6** shows a block diagram of the applied ANNs prediction model. The ANNs predict date quality parameters of the pH, total soluble solids (TSS), water activity (aw), and moisture content (MC) of date fruits during cold storage, which is based on 14 electrical parameters of the capacitance value at the series equivalent circuit model (Cs, nF), the dissipation factor (D), the equivalent series resistance (Rs, kΩ), the equivalent parallel resistance (Rp, kΩ), the capacitance value at the parallel equivalent circuit model (Cp, nF), the inductance value in the series equivalent circuit model(Ls, H), the inductance value in the parallel equivalent circuit model (Lp, H), the resistance (R, kΩ), the direct current resistance (DCR, kΩ), the reactance (X, kΩ), the absolute value of the impedance (Z, kΩ), the phase angle (θ°, degree), the phase radian (θ, rad), and the quality factor (Q).

Srinivasagan et al. [156] used a TinyML-based multispectral sensor for the shelf life estimation of fresh date fruits packed under modified atmospheres. This sensor uses ML algorithms to estimate the shelf life of fresh dates based on various fruit properties such as moisture content, total soluble solids, tannin content, and sugar content. **Figure 7** shows a block diagram of the applied ANNs prediction model. The ANNs predict shelf life and fruit quality based on 18 spectroscopy reflectance, packaging types, and storage temperatures. The ANN used in the study has three layers: an input layer with 20 nodes, two hidden layers, and an output layer with one node. The input layer collects the AS7265x Triad optical sensor data *via* the I2C port of the Arduino Nano33 BLE Sense microcontroller. The hidden layers perform data transformation and feature extraction. The output layer is a single neuron that produces a continuous output value. The ANN layers are designed to predict the shelf life of fresh date fruits based on various fruit properties such as color, texture, and temperature. The authors conducted experiments to validate the sensor's accuracy and found that it could accurately estimate the shelf life of fresh dates. This technology can potentially improve the efficiency of the date fruit industry by reducing waste and increasing

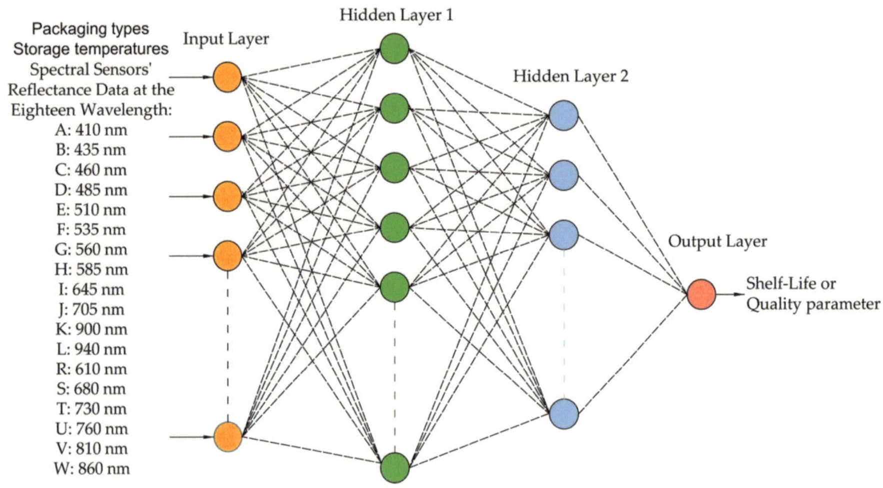

Figure 7.
A simple block diagram of the applied ANNs model for estimating shelf life and quality of date fruit using multispectral sensor during storage under modified atmospheres.

profits. The study also found that modified atmosphere packaging (MAP) can extend the shelf life of date fruits. The researchers observed variations in fresh fruit shelf life estimations across the three main stages of fruit maturity from the Khalal to the Tamr stage. The study suggests that using TinyML for shelf life estimation can improve the efficiency of the date fruit industry by reducing waste and increasing profits. Overall, the study provides a promising approach for estimating the shelf life of fresh date fruits using TinyML and low-cost sensors [156, 157].

5.5 Pest management

One of the most critical problems of date palm mite control is an objective decision-making method for monitoring and predicting date palm mite infestation on date fruits. Mohammed et al. developed, evaluated, and validated prediction models for date palm mite infestation on fruits based on meteorological parameters, i.e., relative humidity, temperature, solar radiation, and wind speed and the physicochemical parameters of date fruits, i.e., firmness, weight, moisture content, total sugar, total soluble solids, and tannin content, using two ML models, i.e., linear regression and decision forest regression. The study is a good example of how ML can improve agricultural practices. The study found that the decision forest regression model was more accurate than the linear regression model in predicting the date palm mite based on the input parameters. This indicates that the decision forest regression model can be used to consider several factors that can affect the infestation of date fruits by the date palm mite. The study's results have important implications for date palm cultivation. The study also found that the developed model could predict the date palm mite count on date palm fruits based on the combination of meteorological and physicochemical properties variables. Farmers can take preventive measures to protect their crops by using ML to predict the date palm mite infestation. This can help to reduce the impact of the date palm mite on the date palm industry and ensure the availability of high-quality dates [66]. **Figure 8** illustrates an experiment to predict date pam infestation on date palm fruits based

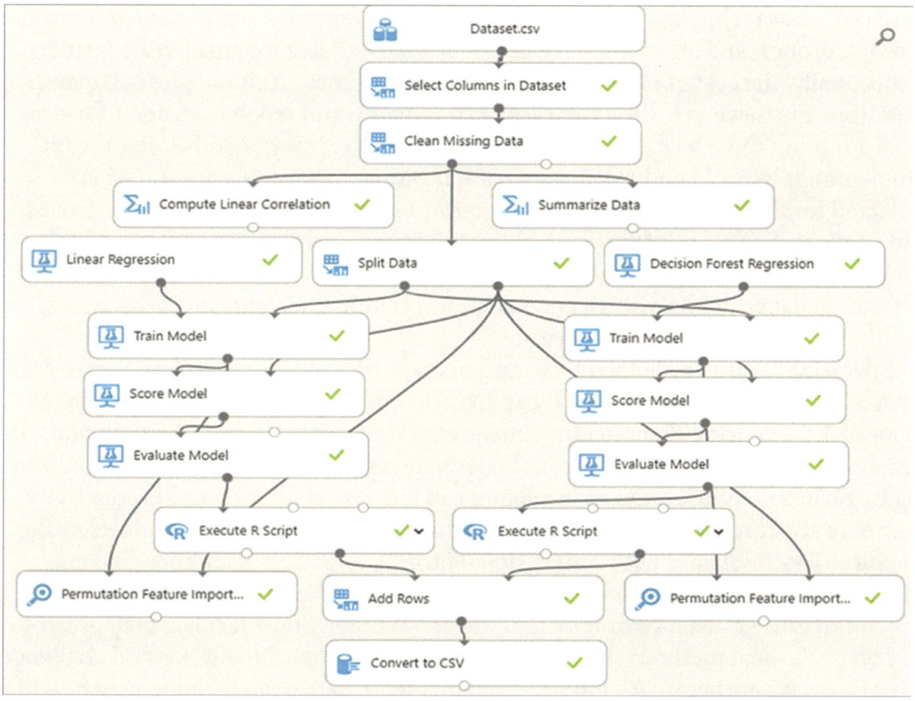

Figure 8.
A screenshot for the architecture of the developed models using Microsoft Azure Machine Learning.

on the study area's meteorological parameters data, the date fruits' physicochemical parameters data during the development stages, and the combined data of meteorological variables and physicochemical properties.

6. Challenges of implementing AI and IoT in date palm cultivation

The transformation brought about by integrating AI and IoT is undeniable, significantly revolutionizing the cultivation of date palms. Farmers can optimize operations, reduce costs, and improve yields using AI and IoT technologies. However, there are several challenges to implementing AI and IoT systems in date palm cultivation. The main challenges to using AI and IoT technologies in date palm cultivation are the lack of infrastructure and connectivity in rural areas, where most farms are located. There is a lack of the necessary infrastructure to support these technologies in many regions, where date palms are grown. For instance, there might not be a stable power source or internet connection to power the devices needed for AI and IoT implementation. This makes it challenging for farmers to gain access to and efficiently utilize these techniques. IoT devices require a stable internet connection to function effectively, and in areas with poor connectivity, the data collected may be inaccurate or incomplete. This can lead to incorrect decision-making and ultimately impact date palm yields and profitability. The high cost of implementing AI and IoT technologies is another challenge.

Implementing IoT and AI systems demands a substantial hardware, software, and training investment. Many small-scale farmers may lack the funding necessary to purchase these technologies. The initial investment required to purchase and install sensors, drones, and other IoT devices can be unaffordable for small-scale farmers. Additionally, the cost of maintaining and upgrading these technologies can increase over time. Furthermore, the expense of maintenance and repairs can deter farmers from adopting these technologies. Data security and privacy are other significant problems applying AI and IoT in date palm farming. Large amounts of data are collected for these technologies from numerous sources, including sensors, drones, and cameras. Sensitive information about the farm's operations is contained in this data, making it vulnerable to misuse or cyberattacks. This is particularly important as farming data can be sensitive and valuable, providing insights into crop yields, soil health, weather patterns, and more.

Special skills are needed to install and manage AI and IoT systems on farms. Many farmers may not have the technical expertise to install and maintain these systems themselves, requiring them to hire outside experts or invest in training their staff. It is also a concern that AI and IoT technologies may replace human labor on farms, leading to job losses. While these technologies can increase efficiency and productivity, they may also lead to a decrease in demand for manual labor. Moreover, integrating AI and IoT technologies with conventional farming practices is another challenge. Many farmers may have little or no prior experience utilizing these technologies, which can cause resistance or reluctance to adopt them. Some farmers may prefer to rely on traditional methods they have used for generations. Finally, several challenges must be overcome before AI and IoT technologies in date palm farming may be widely adopted. Limitations in the infrastructure, high implementation costs, security and data privacy concerns, and reluctance to change are some of these challenges.

7. Future opportunities for AI and IoT in date palm cultivation

Advances in AI and IoT technologies offer new opportunities to improve the efficiency and sustainability of date palm farming. One potential application of AI in date palm cultivation is predictive analytics. By analyzing weather patterns, soil conditions, and other environmental factors, AI algorithms can help farmers predict when to propagate offshoots, irrigate, and harvest their crops for optimal yields. This can help reduce farm waste, increase crop productivity, and improve profitability. Another potential application of AI is precision agriculture. By using sensors and other IoT devices to collect data on soil moisture, temperature, and other environmental factors, farmers can use AI algorithms to optimize water amounts, irrigation schedules, and optimum doses of fertilizer applications. This can help reduce water usage and minimize environmental impact while maximizing palm yields. AI and IoT can also be used to monitor the health of date palm trees. AI algorithms can detect early signs of disease or pest infestations by analyzing images of leaves and trunks. AI noses or electronic noses can diagnose insect pests and diseases, which is a fast and noninvasive approach. This can help farmers take corrective action before the problem spreads, reducing yield losses and minimizing the need for pesticides and fungicides. In addition to AI, IoT technologies such as drones and robots offer new opportunities for precision agriculture in date palm cultivation. Drones equipped with cameras and sensors can provide high-resolution images of date palms, allowing farmers to monitor their health and growth more closely.

Meanwhile, robots can be used for tasks such as pruning and harvesting, reducing labor costs and increasing efficiency. AI and IoT technologies can be used for fruit sorting purposes based on fruit size, weight, color, maturity indices, etc., saving time, labor, and resources. Uneven in situ fruit ripening is a common problem in date palms. AI and IoT technologies can also sort out ripe and unripe fruits. Then the unripe fruits can be ripened artificially using the same technologies to avoid wastage and to enhance farm income.

8. Conclusion

The application of AI and IoT in date palm cultivation in Saudi Arabia shows enormous potential for enhancing productivity, sustainability, and quality management. There is a growing interest in applying AI and IoT in cultivating and managing date palm trees in Saudi Arabia. Several studies have explored AI and IoT technologies in various aspects of date palm cultivation, including pest management, postharvest quality management, yield improvement, and mapping of date palm trees. Using AI and IoT technologies in date palm cultivation offers benefits such as improved crop yield, efficient pest management, and enhanced postharvest quality control. These technologies enable real-time monitoring and data analysis, which can help farmers make informed decisions and optimize resource allocation. Integrating AI and IoT can contribute to the sustainability of date palm cultivation by reducing water consumption and mitigating the impact of climate change. However, it is essential to note that while AI and IoT technologies hold promise, challenges still need to be addressed. These include the need for robust data collection and analysis, ensuring data security and privacy, and addressing the digital divide in rural areas. Further research and development are required to fully harness the potential of AI and IoT in date palm cultivation. Continued research and innovation in this field can contribute to advancing date palm cultivation practices and support the agricultural sector.

Acknowledgements

The authors gratefully acknowledge the financial support from Date Palm Research Center of Excellence, King Faisal University, Saudi Arabia.

Conflict of interest

The authors declare no conflict of interest.

Author details

Maged Mohammed[1,2*], Nashi K. Alqahtani[1,3], Muhammad Munir[1]
and Mohamed A. Eltawil[4,5]

1 Date Palm Research Center of Excellence, King Faisal University, Al-Ahsa, Saudi Arabia

2 Faculty of Agriculture, Agricultural and Biosystems Engineering Department, Menoufia University, Shebin El Koum, Egypt

3 Department of Food and Nutrition Sciences, College of Agricultural and Food Sciences, King Faisal University, Al-Ahsa, Saudi Arabia

4 Department of Agricultural Systems Engineering, College of Agricultural and Food Sciences, King Faisal University, Al-Ahsa, Saudi Arabia

5 Faculty of Agriculture, Agricultural Engineering Department, Kafrelsheikh University, Kafr ElSheikh, Egypt

*Address all correspondence to: memohammed@kfu.edu.sa

References

[1] Edan Y, Han S, Kondo N. Automation in agriculture. In: Nof S, editor. Springer Handbook of Automation. Berlin, Heidelberg: Springer Berlin Heidelberg; 2009. pp. 1095-1128. DOI: 10.1007/978-3-540-78831-7_63

[2] Elhassan Ahmed OM, Osman AA, Awadalkarim SD. A design of an automated fertigation system using IoT. In: 2018 International Conference on Computer, Control, Electrical, and Electronics Engineering (ICCCEEE). Khartoum, Sudan: IEEE; 2018. pp. 1-5. Available from: https://ieeexplore.ieee.org/document/8515772/

[3] Lavanya G, Rani C, Ganeshkumar P. An automated low cost IoT based fertilizer intimation system for smart agriculture. Sustainable Computing: Informatics and Systems. 2020;**28**:100300. Available from: https://linkinghub.elsevier.com/retrieve/pii/S2210537918302567

[4] Ben Ayed R, Hanana M. Artificial intelligence to improve the food and agriculture sector. Khan R, editor. Journal of Food Quality. 2021;**22**:1-7. Available from: https://www.hindawi.com/journals/jfq/2021/5584754/

[5] Boursianis AD, Papadopoulou MS, Diamantoulakis P, Liopa-Tsakalidi A, Barouchas P, Salahas G, et al. Internet of things (IoT) and agricultural unmanned aerial vehicles (UAVs) in smart farming: A comprehensive review. Internet of Things. 2022;**18**:100187. Available from: https://linkinghub.elsevier.com/retrieve/pii/S2542660520300238

[6] Sharma A, Georgi M, Tregubenko M, Tselykh A, Tselykh A. Enabling smart agriculture by implementing artificial intelligence and embedded sensing. Computers and Industrial Engineering. 2022;**165**:107936. Available from: https://linkinghub.elsevier.com/retrieve/pii/S0360835222000067

[7] Jha K, Doshi A, Patel P, Shah M. A comprehensive review on automation in agriculture using artificial intelligence. Artificial Intelligence in Agriculture. 2019;**2**:1-12. Available from: https://linkinghub.elsevier.com/retrieve/pii/S2589721719300182

[8] Suleimenov IE, Vitulyova YS, Bakirov AS, Gabrielyan OA. Artificial intelligence. In: Proceedings of the 2020 6th International Conference on Computer and Technology Applications. New York, NY, USA: ACM; 2020. pp. 22-25. DOI: 10.1145/3397125.3397141

[9] Neethirajan S. The role of sensors, big data and machine learning in modern animal farming. Sensing and Bio-Sensing Research. 2020;**29**:100367. Available from: https://linkinghub.elsevier.com/retrieve/pii/S2214180420301343

[10] Liu SY. Artificial intelligence (AI) in agriculture. IT Professional. 2020;**22**(3):14-15

[11] Linaza MT, Posada J, Bund J, Eisert P, Quartulli M, Döllner J, et al. Data-driven artificial intelligence applications for sustainable precision agriculture. Agronomy. 2021;**11**(6):1227. DOI: 2073-4395/11/6/1227

[12] Basri R, Islam F, Shorif SB, Uddin MS. Robots and drones in agriculture—A survey. In: Computer Vision and Machine Learning in Agriculture. Singapore: Springer; 2021. pp. 9-29. DOI: 10.1007/978-981-33-6424-0_2

[13] Bogue R. Robots poised to transform agriculture. The International Journal of Robotics Research and Application. 2021;**48**(5):637-642

[14] Neethirajan S. Affective state recognition in livestock—Artificial intelligence approaches. Animals. 2022;**12**(6):759. DOI: 2076-2615/12/6/759

[15] Wakchaure M, Patle BK, Mahindrakar AK. Application of AI techniques and robotics in agriculture: A review. Artificial Intelligence in the Life Sciences. 2023;**3**:100057. Available from: https://linkinghub.elsevier.com/retrieve/pii/S2667318523000016

[16] Kim WS, Lee WS, Kim YJ. A review of the applications of the internet of things (IoT) for agricultural automation. Journal of Biosystems Engineering. 2020;**45**(4):385-400

[17] Parvez B, Haidri RA, Kumar VJ. IoT in agriculture. In: 2020 International Conference on Computational Performance Evaluation (ComPE). Shillong, India: IEEE; 2020. pp. 844-847. Available from: https://ieeexplore.ieee.org/document/9200035/

[18] Vangala A, Das AK, Kumar N, Alazab M. Smart secure sensing for IoT-based agriculture: Blockchain perspective. IEEE Sensors Journal. 2021;**21**(16):17591-17607

[19] Sagheer A, Mohammed M, Riad K, Alhajhoj M. A cloud-based IoT platform for precision control of soilless greenhouse cultivation. Sensors (Switzerland). 2021;**21**(1):1-29. DOI: 1424-8220/21/1/223

[20] Mohammed M, Riad K, Alqahtani N. Efficient iot-based control for a smart subsurface irrigation system to enhance irrigation management of date palm. Sensors. 2021;**21**(12):3942. DOI: 10.3390/s21123942

[21] Kashyap PK, Kumar S, Jaiswal A, Prasad M, Gandomi AH. Towards precision agriculture: IoT-enabled intelligent irrigation systems using deep learning neural network. IEEE Sensors Journal. 2021;**21**(16):17479-17491. Available from: https://ieeexplore.ieee.org/document/9388691/

[22] Kailasam S, Achanta SDM, Rama Koteswara Rao P, Vatambeti R, Kayam S. An IoT-based agriculture maintenance using pervasive computing with machine learning technique. International Journal of Intelligent Computing and Cybernetics. 2022;**15**(2):184-197

[23] Zhao J-c, Zhang J-f, Yu F, Guo J-x, Zhao JC, Zhang JF, et al. The study and application of the IOT technology in agriculture. In: 2010 3rd International Conference on Computer Science and Information Technology. Chengdu, China: IEEE; 2010. pp. 462-465. Available from: http://ieeexplore.ieee.org/document/5565120/

[24] Patil VC, Al-Gaadi KA, Biradar DP, Rangaswamy M. Internet of things (Iot) and cloud computing for agriculture: An overview. In: Proceedings of Agro-Informatics and Precision Agriculture (AIPA 2012), India. 1 Aug 2012. pp. 292-296

[25] Muangprathub J, Boonnam N, Kajornkasirat S, Lekbangpong N, Wanichsombat A, Nillaor P. IoT and agriculture data analysis for smart farm. Computers and Electronics in Agriculture. 2019;**156**:467-474. Available from: https://www.sciencedirect.com/science/article/abs/pii/S0168169918308913

[26] Salam A. Internet of things for sustainable community development:

Introduction and overview. In: Salam A, editor. Internet of Things. Cham, Switzerland: Springer International Publishing; 2020. pp. 1-31. DOI: 10.1007/978-3-030-35291-2_1

[27] Rehman A, Saba T, Kashif M, Fati SM, Bahaj SA, Chaudhry H. A revisit of internet of things technologies for monitoring and control strategies in smart agriculture. Agronomy. 2022;**12**(1):127. DOI: 2073-4395/12/1/127

[28] Zaid A, De Wet PF. Date palm cultivation. In: FAO Plant Production and Protection Paper 156 Rev1. Rome, Italy: FAO; 2002. pp. 1-28

[29] Mohammed M, Hamdoun H, Sagheer A. Toward sustainable farming: Implementing artificial intelligence to predict optimum water and energy requirements for sensor-based micro irrigation systems powered by solar PV. Agronomy. 2023;**13**(4):1081. DOI: 2073-4395/13/4/1081

[30] Mohammed M, Riad K, Alqahtani N. Design of a smart IoT-based control system for remotely managing cold storage facilities. Sensors. 2022;**22**(13):4680. DOI: 1424-8220/22/13/4680

[31] Gibril MBA, Shafri HZM, Shanableh A, Al-Ruzouq R, Wayayok A, Hashim SJ. Deep convolutional neural network for large-scale date palm tree mapping from UAV-based images. Remote Sensing. 2021;**13**(14):2787. DOI: 2072-4292/13/14/2787

[32] Dehghanisanij H, Salamati N, Emami S, Emami H, Fujimaki H. An intelligent approach to improve date palm crop yield and water productivity under different irrigation and climate scenarios. Applied Water Science. 2023;**13**(2):56. DOI: 10.1007/s13201-022-01836-8

[33] Ramakrishnam Raju SVS, Dappuri B, Ravi Kiran Varma P, Yachamaneni M, DMG V, Mishra MK. Design and implementation of smart hydroponics farming using IoT-based AI controller with mobile application system. Chelladurai SJS, editor. Journal of Nanomaterials. 2022;**2022**:1-12. Available from: https://www.hindawi.com/journals/jnm/2022/4435591/

[34] Canlas FQ, Al Falahi M, Nair S. IoT based date palm water management system using case-based reasoning and linear regression for trend analysis. International Journal of Advanced Computer Science and Applications. 2022;**13**(2):549-556

[35] Dhehibi B, Ben SM, Frija A, Aw-Hassan A, El OH, Al RYM. Economic and technical evaluation of different irrigation systems for date palm farming system in the GCC countries: Case of Oman. Environment and Natural Resources Research. 2018;**8**(3):55

[36] FAOSTAT. Food and Agriculture Organization of the United Nations. 2021. pp. 403-403. Available from: https://www.fao.org/faostat/en/#data/QCL [Accessed: 28 April 2023]

[37] Ahmed Mohammed ME, Refdan Alhajhoj M, Ali-Dinar HM, Munir M. Impact of a novel water-saving subsurface irrigation system on water productivity, photosynthetic characteristics, yield, and fruit quality of date palm under arid conditions. Agronomy. 2020;**10**(9):1265. DOI: 2073-4395/10/9/1265

[38] Mohammed M, Sallam A, Munir M, Ali-Dinar H. Effects of deficit irrigation scheduling on water use, gas exchange, yield, and fruit quality of date palm. Agronomy. 2021;**11**(11):2256

[39] El Hadrami A, Al-Khayri JM. Socioeconomic and traditional importance of date palm. Emirates Journal of Food and Agriculture. 2012;**24**(5):371-385

[40] Arias E, Hodder AJ, Oihabi A, Jimnez E, Hodder AJ, Oihabi A. FAO support to date palm development around the world: 70 years of activity. Emirates Journal of Food and Agriculture. 2016;**28**(1):1-11. Available from: http://www.ejmanager.com/fulltextpdf.php?mno=204338

[41] Ali-Dinar H, Mohammed M, Munir M. Effects of pollination interventions, plant age and source on hormonal patterns and fruit set of date palm (*Phoenix dactylifera* L.). Horticulturae. 2021;**7**(11):427. DOI: 2311-7524/7/11/427

[42] Alnaim M, Mohamed M, Mohammed M, Munir M. Effects of automated irrigation systems and water regimes on soil properties, water productivity, yield and fruit quality of date palm. Agriculture. 2022;**12**(3):343. DOI: 2077-0472/12/3/343

[43] Mohammed M, Munir M, Ghazzawy HS. Design and evaluation of a smart ex vitro acclimatization system for tissue culture plantlets. Agronomy. 2022;**13**(1):78

[44] Statista. Leading Fruit Dates Exporters Worldwide in 2021. New York, USA: Statista Research Department. Statista Inc.; 2021. Available from: https://www.statista.com/statistics/961359/global-leading-exporters-of-dates

[45] Erskine W, Moustafa AT, Osman AE, Lashine Z, Nejatian A, Badawi T, et al. Date palm in the GCC countries of the Arabian peninsula. In: Proc Regional Workshop on Date Palm Development in the Arabian Peninsula. Abu Dhabi: Icarda; 2004. pp. 29-31. Available from: http://www.icarda.org/aprp/datepalm/introduction/introduction.htm

[46] Tripler E, Shani U, Mualem Y, Ben-Gal A. Long-term growth, water consumption and yield of date palm as a function of salinity. Agricultural Water Management. 2011;**99**(1):128-134

[47] Abul-Soad AA, Jain SM, Jatoi MA. Biodiversity and conservation of date palm. In: Biodiversity and Conservation of Woody Plants, Sustainable Development and Biodiversity. Cham, Switzerland: Springer; 2017. pp. 313-353. DOI: 10.1007/978-3-319-66426-2_12

[48] Chowdhury S, Al-Zahrani M. Characterizing water resources and trends of sector wise water consumptions in Saudi Arabia. Journal of King Saud University - Engineering Sciences. 2015;**27**(1):68-82. Available from: https://linkinghub.elsevier.com/retrieve/pii/S1018363913000081

[49] DeNicola E, Aburizaiza OS, Siddique A, Khwaja H, Carpenter DO. Climate change and water scarcity: The case of Saudi Arabia. Annals of Global Health. 2015;**81**(3):342-353

[50] Baig MB, Alotibi Y, Straquadine GS, Alataway A. Water resources in the kingdom of Saudi Arabia: Challenges and strategies for improvement. Global Issues in Water Policy. 2020;**23**:135-160

[51] Barth HJ. Desertification in the Eastern Province of Saudi Arabia. Journal of Arid Environments. 1999;**43**(4):399-410

[52] Amin AA. The extent of desertification on Saudi Arabia. Environmental Geology. 2004;**46**(1):22-31

[53] El Hassan IM. Desertification Monitoring Using Remote Sensing Technology. In: International Conference on Water Resources & Arid Environment. 2004. pp. 1-15. Available from: http://www.psipw.org/English_PDF/3_Distance/E3-3.pdf

[54] Ejaz N, Elhag M, Bahrawi J, Zhang L, Gabriel HF, Rahman KU. Soil erosion modelling and accumulation using RUSLE and remote sensing techniques: Case study Wadi Baysh, Kingdom of Saudi Arabia. Sustainability. 2023;**15**(4):3218. DOI: 2071-1050/15/4/3218

[55] Elhag M, Hidayatulloh A, Bahrawi J, Chaabani A, Budiman J. Using inconsistencies of Wadi morphometric parameters to understand patterns of soil erosion. Arabian Journal of Geosciences. 2022;**15**(14):1299. Available from: https://link.springer.com/10.1007/s12517-022-10422-w

[56] Salih A, Hassaballa AA, Ganawa E. Mapping desertification degree and assessing its severity in Al-Ahsa oasis, Saudi Arabia, using remote sensing-based indicators. Arabian Journal of Geosciences. 2021;**14**(3):192. DOI: 10.1007/s12517-021-06523-7

[57] Mohammed M, Elmahmoudi A, Almolhem Y. Applications of electromagnetic induction and electrical resistivity tomography for digital monitoring and assessment of the soil: A case study of Al-Ahsa Oasis, Saudi Arabia. Applied Sciences. 2022;**12**(4):2067. DOI: 2076-3417/12/4/2067

[58] El-Shafie HAF, Faleiro JR, Al-Abbad AH, Stoltman L, Mafra-Neto A. Bait-free attract and kill technology (hook™ RPW) to suppress red palm weevil, rhynchophorus ferrugineus (coleoptera: Curculionidae) in date palm. Florida Entomologist. 2011;**94**(4):774-778

[59] El-Shafie HAHA, Mohammed MEME, Sallam AKAKA. Quarantine protocol against coleopteran borers in date palm offshoots using eco2fume gas. Outlooks on Pest Management. 2020;**31**(4):190-192

[60] Mohammed M, El-shafie H, Alqahtani N, El-Shafie H, Alqahtani N. Design and validation of computerized flight-testing systems with controlled atmosphere for studying flight behavior of red palm weevil, *Rhynchophorus ferrugineus* (Olivier). Sensors. 2021;**21**(6):2112. DOI: 1424-8220/21/6/2112

[61] El-Shafie HAFAF, Abdel-Banat BMAMA, Mohammed MEAEA, Al-Hajhoj MRR. Monitoring tools and sampling methods for major date palm pests. CAB Reviews: Perspectives in Agriculture, Veterinary Science, Nutrition and Natural Resources. 2019;**14**(022):1-11

[62] Mohammed MEA, El-Shafie HA, Sallam AAA. A solar-powered heat system for management of almond moth, *Cadra cautella* (Lepidoptera: Pyralidae) in stored dates. Postharvest Biology and Technology. 2019;**154**(March):121-128

[63] Mohammed MEA, El-Shafie HAF, Alhajhoj MR. Recent trends in the early detection of the invasive red palm weevil, *Rhynchophorus ferrugineus* (Olivier). In: Invasive Species—Introduction Pathways, Economic Impact, and Possible Management Options. London, UK: IntechOpen; 2020

[64] Mohammed MEAMEA, El-Shafie HAFHAF, Al-Hajhoj MBRMBR. Design of an automated solar-powered light trap for monitoring and mass trapping of major date palm pests.

Ecology, Environment and Conservation. 2018;**24**(1):177-185

[65] El-Shafie H, Mohammed M, Alqahtani N. A preliminary study on flight characteristics of the longhorn date palm stem Borer Jebusaea hammerschmidtii (Reiche 1878) (Coleoptera: Cerambycidae) using a computerized flight mill. Agriculture. 2022;**12**(1):120

[66] Mohammed M, El-Shafie H, Munir M. Development and validation of innovative machine learning models for predicting date palm mite infestation on fruits. Agronomy. 2023;**13**(2):494. DOI: 2073-4395/13/2/494

[67] Sarraf M, Jemni M, Kahramanoğlu I, Artés F, Shahkoomahally S, Namsi A, et al. Commercial techniques for preserving date palm (*Phoenix dactylifera*) fruit quality and safety: A review. Saudi Journal of Biological Sciences. 2021;**28**(8):4408-4420

[68] Kader AA, Hussein AM, Adel AK, Awad MH. Harvesting and postharvest handling of dates. Icarda. 2009;**15**(September):1-15

[69] El-Habbab PMS, Al-Mulhim F, Al-Eid A, El-Saad MA, Aljassas A, Sallam SL, et al. Assessment of post-harvest loss and waste for date palms in the Kingdom of Saudi Arabia. International Journal of Environmental & Agriculture Research. 2017;**3**(6):01-11. Available from: http://ijoear.com/Paper-June-2017/IJOEAR-MAY-2017-7.pdf

[70] Mohammed MEAMEA, El-Shafie HAFHAF, Alhajhoj MRMR. Design and efficacy evaluation of a modern automated controlled atmosphere system for pest management in stored dates. Journal of Stored Products Research. 2020;**89**:101719. Available from: https://www.sciencedirect.com/science/article/pii/S0022474X20304641

[71] Mohammed M, Alqahtani N, El-Shafie H. Development and evaluation of an ultrasonic humidifier to control humidity in a cold storage room for postharvest quality management of dates. Food. 2021;**10**(5):949. DOI: 2304-8158/10/5/949

[72] Almutawa AA. Date production in the Al-Hassa region, Saudi Arabia in the face of climate change. Journal of Water and Climate Change. 2022;**13**(7):2627-2647. Available from: https://iwaponline.com/jwcc/article/13/7/2627/89124/Date-production-in-the-Al-Hassa-region-Saudi

[73] Aleid SM, Al-Khayri JM, Al-Bahrany AM. Date palm status and perspective in Saudi Arabia. In: Date Palm Genetic Resources and Utilization. Dordrecht: Springer Netherlands; 2015. pp. 49-95. DOI: 10.1007/978-94-017-9707-8_3

[74] Ploennigs J, Cohn J, Stanford-Clark A. The future of IoT. IEEE Internet of Things Magazine. 2018;**1**(1):28-33

[75] Osifeko MO, Hancke GP, Abu-Mahfouz AM. Artificial intelligence techniques for cognitive sensing in future IoT: State-of-the-art, potentials, and challenges. Journal of Sensor and Actuator Networks. 2020;**9**(2):21. DOI: 2224-2708/9/2/21

[76] Lin ZX, Ying ZM, Bing LQ, Chao CC, Chao TY. FinBrain: When finance meets AI 2.0. Frontiers of Information Technology & Electronic Engineering. 2019;**20**(7):914-924

[77] Mohamed E. The relation of artificial intelligence with internet of things: A survey. Journal of

Cybersecurity and Information Management. 2020;**1**(1):30-34

[78] Mohammed M, Munir M, Aljabr A. Prediction of date fruit quality attributes during cold storage based on their electrical properties using artificial neural networks models. Food. 2022;**11**(11):1666. DOI: 2304-8158/11/11/1666

[79] Gopalan SS, Raza A, Almobaideen W. IoT security in healthcare using AI: A survey. In: 2020 International Conference on Communications, Signal Processing, and their Applications (ICCSPA). Sharjah, United Arab Emirates: IEEE; 2021. pp. 1-6. Available from: https://ieeexplore.ieee.org/document/9385711/

[80] Milić DC, Tolić IH, Peko M. Internet of things (Iot) solutions in smart transportation management. Business Logistics in Modern Management. 2020;**20**:331-343

[81] Misra NN, Dixit Y, Al-Mallahi A, Bhullar MS, Upadhyay R, Martynenko A. IoT, big data, and artificial intelligence in agriculture and food industry. IEEE Internet of Things Journal. 2022;**9**(9):6305-6324

[82] Prasad M, Prasad R, Kulkarni U. A decision support system for agriculture using natural language processing (ADSS). Proceedings of the International MultiConference of Engineers and Computer Science. 2008;**I**:19-21

[83] Devi M, Dua M. ADANS: An agriculture domain question answering system using ontologies. In: 2017 International Conference on Computing, Communication and Automation (ICCCA). Greater Noida, India: IEEE; 2017. pp. 122-127. Available from: http://ieeexplore.ieee.org/document/8229784/

[84] Gunawan R, Taufik I, Mulyana E, Kurahman OT, Ramdhani MA, Mahmud. Chatbot application on internet of things (IoT) to support smart urban agriculture. In: 2019 IEEE 5th International Conference on Wireless and Telematics (ICWT). Yogyakarta, Indonesia: IEEE; 2019. pp. 1-6. Available from: https://ieeexplore.ieee.org/document/8978223/

[85] Yunpeng C, Jian W, Juan L. The development of deep learning based natural language processing (NLP) technology and applications in agriculture. Journal of Agricultural Big Data. 2019;**1**(1):38-44. Available from: https://www.cabdirect.org/cabdirect/abstract/20219909618

[86] Zhang Q, Liu Y, Gong C, Chen Y, Yu H. Applications of deep learning for dense scenes analysis in agriculture: A review. Sensors. 2020;**20**(5):1520. DOI: 1424-8220/20/5/1520

[87] Gomes JFS, Leta FR. Applications of computer vision techniques in the agriculture and food industry: A review. European Food Research and Technology. 2012;**235**(6):989-1000

[88] Patrício DI, Rieder R. Computer vision and artificial intelligence in precision agriculture for grain crops: A systematic review. Computers and Electronics in Agriculture. 2018;**153**:69-81. Available from: https://linkinghub.elsevier.com/retrieve/pii/S0168169918305829

[89] Li Y, Zhang Y. Application research of computer vision Technology in Automation. In: 2020 International Conference on Computer Information and Big Data Applications (CIBDA). Guiyang, China: IEEE; 2020. pp. 374-377. Available from: https://ieeexplore.ieee.org/document/9148583/

[90] Tian H, Wang T, Liu Y, Qiao X, Li Y. Computer vision technology in agricultural automation—A review.

Information Processing in Agriculture. 2020;**7**(1):1-19

[91] Tripathi MK, Maktedar DD. A role of computer vision in fruits and vegetables among various horticulture products of agriculture fields: A survey. Information Processing in Agriculture. 2020;**7**(2):183-203

[92] Grift T, Zhang Q, Kondo N, Ting K. A review of automation and robotics for the bio-industry. Journal of Biomechatronics Engineering. 2008;**1**(1):37-54

[93] King A. Technology: The future of agriculture. Nature. 2017;**544**(7651):21-23

[94] Marinoudi V, Sørensen CG, Pearson S, Bochtis D. Robotics and labour in agriculture. A context consideration. Biosystems Engineering. 2019;**184**:111-121. Available from: https://linkinghub.elsevier.com/retrieve/pii/S1537511019303617

[95] Sparrow R, Howard M. Robots in agriculture: Prospects, impacts, ethics, and policy. Precision Agriculture. 2021;**22**(3):818-833

[96] Sivaganesan D. Design and development Ai-enabled edge computing for intelligent-Iot applications. Journal of Trends in Computer Science and Smart Technology. 2019;**1**(02):84-94

[97] Zhang X, Cao Z, Dong W. Overview of edge computing in the agricultural internet of things: Key technologies, applications, challenges. IEEE Access. 2020;**8**:141748-141761. Available from: https://ieeexplore.ieee.org/document/9153160/

[98] Guillén MA, Llanes A, Imbernón B, Martínez-España R, Bueno-Crespo A, Cano JC, et al. Performance evaluation of edge-computing platforms for the prediction of low temperatures in agriculture using deep learning. The Journal of Supercomputing. 2021;**77**(1):818-840

[99] Kalyani Y, Collier R. A systematic survey on the role of cloud, fog, and edge computing combination in smart agriculture. Sensors. 2021;**21**(17):5922. DOI: 1424-8220/21/17/5922

[100] Sharofidinov F, MSA M, Pham VD, Khakimov A, Muthanna A, Samouylov K. Agriculture Management Based on LoRa Edge Computing System. In: Distributed Computer and Communication Networks: 23rd International Conference, DCCN 2020, Moscow, Russia; September 14-18, 2020; Moscow, Russia. Vol. 2020. pp. 113-125

[101] Bourechak A, Zedadra O, Kouahla MN, Guerrieri A, Seridi H, Fortino G. At the confluence of artificial intelligence and edge computing in IoT-based applications: A review and new perspectives. Sensors. 2023;**23**(3):1639. DOI: 1424-8220/23/3/1639

[102] Darwish A, Hassanien AE, Elhoseny M, Sangaiah AK, Muhammad K. The impact of the hybrid platform of internet of things and cloud computing on healthcare systems: Opportunities, challenges, and open problems. Journal of Ambient Intelligence and Humanized Computing. 2019;**10**(10):4151-4166

[103] Gill SS, Tuli S, Xu M, Singh I, Singh KV, Lindsay D, et al. Transformative effects of IoT, blockchain and artificial intelligence on cloud computing: Evolution, vision, trends and open challenges. Internet of Things. 2019;**8**:100118. Available from: https://linkinghub.elsevier.com/retrieve/pii/S2542660519302331

[104] Alshehri F, Muhammad G. A comprehensive survey of the internet

of things (IoT) and AI-based smart healthcare. IEEE Access. 2021;**9**:3660-3678. Available from: https://ieeexplore.ieee.org/document/9311140/

[105] Mustapha UF, Alhassan AW, Jiang DN, Li GL. Sustainable aquaculture development: A review on the roles of cloud computing, internet of things and artificial intelligence (CIA). Reviews in Aquaculture. 2021;**13**(4):2076-2091

[106] Mohammed Sadeeq M, Abdulkareem NM, Zeebaree SRM, Mikaeel Ahmed D, Saifullah Sami A, Zebari RR. IoT and cloud computing issues, challenges and opportunities: A review. Qubahan Academic Journal. 2021;**1**(2):1-7

[107] Atlam HF, Azad MA, Alzahrani AG, Wills G. A review of blockchain in internet of things and Ai. Big Data and Cognitive Computing. 2020;**4**(4):1-27

[108] Mohanta BK, Jena D, Satapathy U, Patnaik S. Survey on IoT security: Challenges and solution using machine learning, artificial intelligence and blockchain technology. Internet of Things. 2020;**11**:100227. Available from: https://linkinghub.elsevier.com/retrieve/pii/S2542660520300603

[109] Sandner P, Gross J, Richter R. Convergence of blockchain, IoT, and AI. Frontiers in Blockchain. 2020;**3**:522600. Available from: https://www.frontiersin.org/article/10.3389/fbloc.2020.522600/full

[110] Hu S, Huang S, Huang J, Su J. Blockchain and edge computing technology enabling organic agricultural supply chain: A framework solution to trust crisis. Computers and Industrial Engineering. 2021;**153**:107079. Available from: https://linkinghub.elsevier.com/retrieve/pii/S036083522030749X

[111] Javaid M, Haleem A, Pratap Singh R, Khan S, Suman R. Blockchain technology applications for industry 4.0: A literature-based review. Blockchain: Research and Applications. 2021;**2**(4):100027. Available from: https://linkinghub.elsevier.com/retrieve/pii/S2096720921000221

[112] Navarro E, Costa N, Pereira A. A systematic review of iot solutions for smart farming. Sensors (Switzerland). 2020;**20**(15):1-29

[113] Khan N, Ray RL, Sargani GR, Ihtisham M, Khayyam M, Ismail S. Current progress and future prospects of agriculture technology: Gateway to sustainable agriculture. Sustainability. 2021;**13**(9):4883. DOI: 2071-1050/13/9/4883

[114] Reddy Maddikunta PK, Hakak S, Alazab M, Bhattacharya S, Gadekallu TR, Khan WZ, et al. Unmanned aerial vehicles in smart agriculture: Applications, requirements, and challenges. IEEE Sensors Journal. 2021;**21**(16):17608-17619

[115] Rejeb A, Rejeb K, Abdollahi A, Al-Turjman F, Treiblmaier H. The interplay between the internet of things and agriculture: A bibliometric analysis and research agenda. Internet of Things. 2022;**19**:100580. Available from: https://linkinghub.elsevier.com/retrieve/pii/S2542660522000701

[116] Marline Joys Kumari N, Thirupathi Rao N, Bhattacharyya D. Smart irrigation and cultivation recommendation system for precision agriculture driven by IoT. In: Machine Intelligence, Big Data Analytics, and IoT in Image Processing. Beverly, MA, United States: Wiley; 2023. pp. 123-149. DOI: 10.1002/9781119865513.ch6

[117] Harjeet K, Prashar D. Machine vision technology, deep learning,

and IoT in agricultural industry. In: Industrial Internet of Things. Boca Raton: CRC Press; 2022. pp. 143-159. Available from: https://www.taylorfrancis.com/books/9781003102267/chapters/10.1201/9781003102267-8

[118] Akkem Y, Biswas SK, Varanasi A. Smart farming using artificial intelligence: A review. Engineering Applications of Artificial Intelligence. 2023;**120**:105899

[119] Javaid M, Haleem A, Khan IH, Suman R. Understanding the potential applications of artificial intelligence in agriculture sector. Advanced Agrochem. 2023;**2**(1):15-30. Available from: https://linkinghub.elsevier.com/retrieve/pii/S277323712200020X

[120] Mohan DS, Dhote V, Mishra P, Singh P, Srivastav A. IoT framework for precision agriculture: Machine learning crop prediction. International Journal of Intelligent Systems and Applications in Engineering. 2023;**11**(5s):300-313

[121] Alreshidi E. Smart sustainable agriculture (SSA) solution underpinned by internet of things (IoT) and artificial intelligence (AI). International Journal of Advanced Computer Science and Applications. 2019;**10**(5):93-102

[122] Delgado JA, Short Jr NM, Roberts DP, Vandenberg B. Big data analysis for sustainable agriculture on a geospatial cloud framework. Frontiers in Sustainable Food Systems. 2019;**3**:54. Article 54. Available from: https://www.frontiersin.org/article/10.3389/fsufs.2019.00054/full

[123] Gupta G, Setia R, Meena A, Jaint B. Environment monitoring system for agricultural application using IoT and predicting crop yield using various data mining techniques. In: 2020 5th International Conference on Communication and Electronics Systems (ICCES). Coimbatore, India: IEEE; 2020. pp. 1019-1025. Available from: https://ieeexplore.ieee.org/document/9138032/

[124] Saiz-Rubio V, Rovira-Más F. From smart farming towards agriculture 5.0: A review on crop data management. Agronomy. 2020;**10**(2):207. DOI: 10.3390/agronomy10020207

[125] Singh R, Srivastava S, Mishra R. AI and IoT based monitoring system for increasing the yield in crop production. In: 2020 International Conference on Electrical and Electronics Engineering (ICE3). Gorakhpur, India: IEEE; 2020. pp. 301-305. Available from: https://ieeexplore.ieee.org/document/9122894/

[126] Aggarwal N, Singh D. Technology assisted farming: Implications of IoT and AI. In: IOP Conference Series: Materials Science and Engineering. Vol. 1022, No. 1. IOP Publishing; 2021. p. 012080. DOI: 10.1088/1757-899X/1022/1/012080. Available from: https://iopscience.iop.org/article/10.1088/1757-899X/1022/1/012080/meta

[127] Dahane A, Benameur R, Kechar B. An IoT low-cost smart farming for enhancing irrigation efficiency of smallholders farmers. Wireless Personal Communications. 2022;**127**(4):3173-3210

[128] Raj EFI, Appadurai M, Athiappan K. Precision farming in modern agriculture. In: Smart Agriculture Automation Using Advanced Technologies: Data Analytics and Machine Learning, Cloud Architecture, Automation and IoT. Singapore: Springer; 2022. pp. 61-87. DOI: 10.1007/978-981-16-6124-2_4

[129] Swaminathan B, Palani S, Vairavasundaram S, Kotecha K, Kumar V. IoT-driven artificial intelligence

technique for fertilizer recommendation model. IEEE Consumer Electronics Magazine. 2023;**12**(2):109-117

[130] Widianto MH, Ardimansyah MI, Pohan HI, Hermanus DR. A systematic review of current trends in artificial intelligence for smart farming to enhance crop yield. Journal of Robotics and Control. 2022;**3**(3):269-278

[131] Saranya T, Deisy C, Sridevi S, Anbananthen KSM. A comparative study of deep learning and internet of things for precision agriculture. Engineering Applications of Artificial Intelligence. 2023;**122**:106034. Available from: https://linkinghub.elsevier.com/retrieve/pii/S095219762300218X

[132] Alfian G, Syafrudin M, Fitriyani NL, Rhee J, Ma'arif MR, Riadi I. Traceability system using IoT and forecasting model for food supply chain. In: 2020 International Conference on Decision Aid Sciences and Application (DASA). Sakheer, Bahrain: IEEE; 2020. pp. 903-907. Available from: https://ieeexplore.ieee.org/document/9317011/

[133] Said Mohamed E, Belal AA, Kotb Abd-Elmabod S, El-Shirbeny MA, Gad A, Zahran MB. Smart farming for improving agricultural management. Egyptian Journal of Remote Sensing and Space Sciences. 2021;**24**(3):971-981

[134] Kaur A, Singh G, Kukreja V, Sharma S, Singh S, Yoon B. Adaptation of IoT with blockchain in food supply chain management: An analysis-based review in development, benefits and potential applications. Sensors. 2022;**22**(21):8174. DOI: 1424-8220/22/21/8174

[135] Kutyauripo I, Rushambwa M, Chiwazi L. Artificial intelligence applications in the agrifood sectors. Journal of Agriculture and Food Research. 2023;**11**:100502. Available from: https://linkinghub.elsevier.com/retrieve/pii/S2666154323000091

[136] Alonso RS, Sittón-Candanedo I, García Ó, Prieto J, Rodríguez-González S. An intelligent edge-IoT platform for monitoring livestock and crops in a dairy farming scenario. Ad Hoc Networks. 2020;**98**:102047. Available from: https://linkinghub.elsevier.com/retrieve/pii/S1570870519306043

[137] Ilyas QM, Ahmad M. Smart farming: An enhanced pursuit of sustainable remote livestock tracking and geofencing using IoT and GPRS. Anisi MH, editor. Wireless Communications and Mobile Computing. 2020;**2020**:1-12. Available from: https://www.hindawi.com/journals/wcmc/2020/6660733/

[138] Akhigbe BI, Munir K, Akinade O, Akanbi L, Oyedele LO. IoT technologies for livestock management: A review of present status, opportunities, and future trends. Big Data and Cognitive Computing. 2021;**5**(1):10. DOI: 2504-2289/5/1/10

[139] Alanezi MA, Shahriar MS, Hasan MB, Ahmed S, Sha'aban YA, Bouchekara HREH. Livestock management with unmanned aerial vehicles: A review. IEEE Access. 2022;**10**:45001-45028. Available from: https://ieeexplore.ieee.org/document/9759302/

[140] Farooq MS, Sohail OO, Abid A, Rasheed S. A survey on the role of IoT in agriculture for the implementation of smart livestock environment. IEEE Access. 2022;**10**:9483-9505. Available from: https://ieeexplore.ieee.org/document/9681084/

[141] Mishra S, Sharma SK. Advanced contribution of IoT in agricultural production for the development of smart livestock environments. Internet of

Things. 2023;**22**:100724. Available from: https://linkinghub.elsevier.com/retrieve/pii/S2542660523000471

[142] Alaa H, Waleed K, Samir M, Tarek M, Sobeah H, Salam MA. An intelligent approach for detecting palm trees diseases using image processing and machine learning. International Journal of Advanced Computer Science and Applications. 2020;**11**(7):434-441

[143] Ines N, Hammadi H, Ridha E, Neji I. An intelligent approach to identify the date palm varieties using leaves and fruits an intelligent approach to identify the date palm varieties using leaves and fruits. EasyChair Preprint No. 6323. 17 Aug 2021:6323. Available from: https://easychair.org/publications/preprint/dmwR

[144] Mohammed M, Alqahtani NK. Design and validation of automated sensor-based artificial ripening system combined with ultrasound Pretreatment for date fruits. Agronomy. 2022;**12**(11):2805. DOI: 2073-4395/12/11/2805

[145] Gonzales-Gustavson E, Rusiñol M, Medema G, Calvo M, Girones R. Quantitative risk assessment of norovirus and adenovirus for the use of reclaimed water to irrigate lettuce in Catalonia. Water Research. 2019;**153**:91-99

[146] Eltawil MA, Alhashem HA, Alghannam AO. Design of a solar PV powered variable frequency drive for a bubbler irrigation system in palm trees fields. Process Safety and Environment Protection. 2021;**152**:140-153. Available from: https://linkinghub.elsevier.com/retrieve/pii/S0957582021002792

[147] Wardle K, Dobbs EB, Short KC. In vitro acclimatization of aseptically cultured plantlets to humidity. Journal of the American Society for Horticultural Science. 2022;**108**(3):386-389

[148] Kadleček P, Tichá I, Haisel D, Apková V, Schäfer C. Importance of in vitro pretreatment for ex vitro acclimatization and growth. Plant Science. 2001;**161**(4):695-701

[149] Gao Y, Ji J, Guo Z, Su P. Comparison of the solar PV cooling system and other cooling systems. International Journal of Low-Carbon Technologies. 2018;**13**(4):353-363

[150] Ayadi O, Al-Dahidi S. Comparison of solar thermal and solar electric space heating and cooling systems for buildings in different climatic regions. Solar Energy. 2019;**188**:545-560. Available from: https://linkinghub.elsevier.com/retrieve/pii/S0038092X19306085

[151] Ekren O, Yilanci A, Cetin E, Ozturk HK. Experimental performance evaluation of a PV-powered refrigeration system. Elektron Ir Elektrotechnika. 2011;**8**(114):7-10

[152] Eltawil MA, Mohammed M, Alqahtani NM. Developing machine learning-based intelligent control system for performance optimization of solar PV-powered refrigerators. Sustainability. 2023;**15**(8):6911. DOI: 2071-1050/15/8/6911

[153] Afreen H, Bajwa IS. An IoT-based real-time intelligent monitoring and notification system of cold storage. IEEE Access. 2021;**9**:38236-38253

[154] Chawla R, Sheokand A, Rai MR, Kumar R. Impact of climate change on fruit production and various approaches to mitigate these impacts. The Pharma Innovation Journal. 2021;**10**(3):564-571

[155] Mohammed M, Sallam A, Alqahtani N, Munir M. The combined

effects of precision-controlled temperature and relative humidity on artificial ripening and quality of date fruit. Food. 2021;**10**(11):2636. DOI: 2304-8158/10/11/2636

[156] Srinivasagan R, Mohammed M, Alzahrani A. TinyML-sensor for shelf life estimation of fresh date fruits. Sensors. 2023;**23**(16):7081. DOI: 1424-8220/23/16/7081

[157] Al-Eid SM, Barber AR, Rettke M, Leo A, Alsenaien WA, Sallam AA, et al. Utilisation of modified atmosphere packaging to extend the shelf life of Khalas fresh dates. International Journal of Food Science and Technology. 2012;**47**(7):1518-1525

Chapter 4

Application of Internet of Things (IoT) in Biomedicine: Challenges and Future Directions

Robert Fuior, Alexandru Sălceanu, Cătălina Luca and Călin Corciovă

Abstract

The Internet of Things (IoT) is currently rapidly being incorporated into many fields, but there are still some fields, such as healthcare, where IoT adoption is much slower. Medical IoT refers to a range of medical devices and people that rely on wireless communication to enable healthcare data exchange, remote monitoring, and patient rehabilitation for a better quality of life for the patient. Medical IoT can provide better medical care and rehabilitation services under the careful supervision of the physician, resulting in more cost-effective systems for hospitals as well as for the patient. Due to the regulatory, ethical, and technological challenges of biomedical hardware, the growth of medical IoT is still inhibited. The chapter provides an overview of the various technologies and protocols used for the Internet of Medical Things (IoT), with an overview of the current technologies, applications, and challenges.

Keywords: wearable technology, medical devices, remote patient monitoring, healthcare, protocols

1. Introduction

The Internet of Things (IoT) is a network of individually identifiable connected things (sometimes referred to as devices, objects, and items) that offer services for intelligent computing [1]. IoT objects are also referred to as "smart things," which make it possible to carry out routine tasks in a logical manner. The Internet of Things also benefits human communication in good ways. The Internet of Things (IoT) includes a variety of technologies, such as pervasive computing, sensor technology, embedded systems, communication technologies, sensor networks, Internet protocols, etc., which ultimately support the economic development of contemporary civilizations [2]. IoT's basic tenet is the provision of seamless, all-pervasive connectivity between objects and people.

Given the commonalities between similar technologies and the confluence of three distinct ideas, it is difficult to establish a specific definition of IoT. In a nutshell, IoT is a system in which things are linked in such a way that they may intelligently interact with one another and with people. However, in order to better grasp the Internet of

Things, a number of standard organizations and development groups have established their own definitions [3].

1.1 Evolution of the IoT concept

The evolution of IoT with reference to technological progress in the conception of the Internet is presented in **Figure 1**. The concept of IoT, or the Internet of Things, refers to the interconnectivity of physical devices that are embedded with sensors, software, and network connectivity, allowing them to communicate with each other and with other systems over the Internet [4]. IoT has evolved over the years as a result of advancements in several areas of technology, including embedded systems, M2M communication, CPS, WSN, and the WoT.

These advancements have enabled the development of devices that can collect and transmit data, analyze and respond to that data in real time, and interact with other devices and systems. The development of the Internet itself has also played a critical role in the evolution of IoT. As the Internet has become faster, more reliable, and more widely available, it has made it possible to connect more devices and systems and to transmit larger amounts of data more quickly [5]. Overall, the evolution of IoT has been closely tied to technological progress in a range of areas, and it will likely continue to evolve as new technologies and capabilities become available.

These capabilities will be expanded through interactions with a wide range of electronic devices, according to the concept of the new IoT trend. In general, Internet-centric and Internet-centric items can be thought of as the IoT vision. The improvements of all technologies connected to the idea of "Smart Things" are included in the thing-centric vision. The Internet-centric vision, on the other hand, calls for the development of network technology to connect interactive smart objects with the storage, integration, and management of created data. These perspectives allow the IoT system to be understood as a dynamic dispersed network of intelligent objects that can generate, store, and consume the necessary information [6].

1.2 Application areas of the Internet of Things

Applications based on the Internet of Things (IoT) are expanding quickly, which is causing the world to change [7]. IoT's expansion has been a wonderful development in recent years. IoT is the interconnection of physical and digital items that have been fitted with sensors, software, and other technologies [8]. It entails using the Internet to communicate and exchange data with other systems and devices all around the world. IoT also resembles a group of network-capable gadgets that do not include desktop and laptop computers or servers. IoT has impacted a wide range of industries, starting with the healthcare industry. It is now implantable, wearable, and portable, creating a pervasive and interactive world [9]. It transforms the inanimate objects in

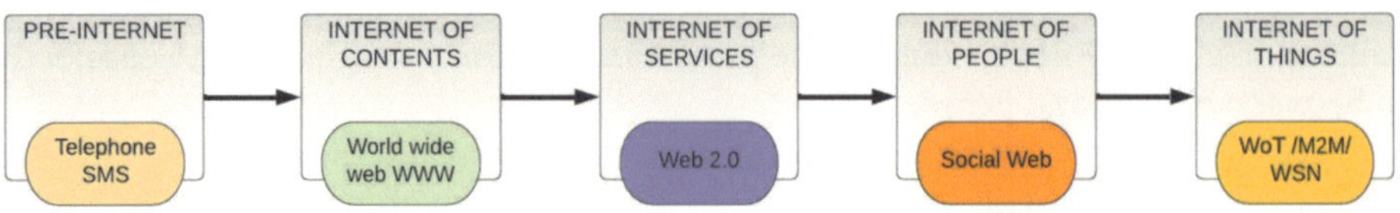

Figure 1.
Technological progression in IoT.

our immediate environment into intelligent objects, resulting in the creation of an information environment that raises the standard of living for people.

IoT devices, for instance, monitor and gather vital measurements (such as blood pressure, blood sugar level, pulse, etc.) in real time, enabling emergency alerts to raise the patient's likelihood of survival [10]. Furthermore, autonomous and self-driving cars assist drivers in getting where they are going by preventing them from going off the road or getting into accidents. Additionally, these definitions are expanded to include automatic emergency alerts for the closest roads as well as medical aid in the event of an accident. The Internet of Things also includes a wide range of contemporary industries, including the manufacturing, assembly, packaging, logistics, smart city, and aviation sectors [11]. **Figure 2** depicts a few of the most important IoT-based application sectors in the fields of health, business, communication, and entertainment.

Internet of Things (IoT) has numerous applications in the medical field, ranging from patient monitoring and diagnosis to drug development and supply chain management. Some of the medical application areas of IoT are:

- Remote patient monitoring: IoT can be used to remotely monitor patients' health conditions, vital signs, and medication adherence, allowing healthcare providers to deliver personalized care and intervene quickly if necessary. This can be

Figure 2.
Important application areas of the Internet of Things.

especially useful for patients with chronic conditions or those who live in remote areas.

- Wearable medical devices: IoT-enabled wearable devices, such as smartwatches, fitness trackers, and health monitors, can track patients' activity levels, sleep patterns, and other health metrics. This information can be used to provide personalized health recommendations and alerts and to help patients manage chronic conditions.

- Connected medical devices: IoT can connect medical devices such as blood glucose meters, blood pressure monitors, and ECG machines, allowing healthcare providers to remotely monitor patient data and provide real-time interventions when necessary.

- Telemedicine: IoT can be used to enable remote consultations between patients and healthcare providers, allowing patients to receive care from anywhere with an Internet connection. This can improve access to care and reduce healthcare costs.

- Drug development: IoT can be used to track drug efficacy and patient outcomes in clinical trials, allowing researchers to optimize drug development and accelerate the drug approval process.

- Supply chain management: IoT can be used to track the temperature, humidity, and other environmental factors that affect the quality and safety of pharmaceuticals and medical supplies during transportation and storage.

Overall, the medical application areas of IoT have the potential to improve patient outcomes, increase efficiency, and reduce costs in the healthcare industry [12–14].

2. IoT in biomedicine: Medical IoT

In the medical field, IoT can be useful in remote patient monitoring (monitoring blood pressure, checking heart rate, checking biometric parameters, or even checking hearing aids), it can be used in the management of diseases in chronic patients or in case of medical emergencies. The advantages of IoT systems used in medicine are that they can continuously and reliably monitor patients and facilitate the digital storage of patients' personal health information. This type of technology helps in the formation of medical databases and their interconnection for a much better management of patient care. Studies show that at this moment, IoT technology will lead to the greatest advances in medicine and will generate revolutionary treatments for patients [15].

IoMT or the Internet of Medical Objects represents the Internet of Things used in healthcare applications. This relatively new market is in continuous growth worldwide with a valuation of over 150 billion dollars in the year 2022 and an increase of $357.45 billion from 2022 to 2028. There are several IoMT solutions on the market but also at the study level that include sensors, wearable systems, including remote access to medical services and monitoring systems of daily activities or hospital systems that increase the quality of patient care, thus fully covering the need for patient care [16].

2.1 Technologies

At this moment in the market, when it comes to healthcare Internet of Things technology, there are several companies that are changing the way IoT is used in medicine. IoMT includes medical devices but with assault and monitoring of public health services, chronic patient care, and distance [17].

In the following lines, we will describe the main IoMT technologies and their principles.

2.1.1 Integrated platforms

Software and hardware integrations for multiple data extractions between medical devices are a real necessity at this point. In this direction, those at Elemental Machines have developed LabOps, a hardware and cloud-based software platform that helps laboratory operations for research and development, clinical laboratories, quality control, and diagnostics of the type [18].

Those from Philips have developed the Phillips Capsule Platform that allows the easy integration of devices used in patient monitoring, thus allowing much easier access to medical data [19].

Also, data about the environment in which we live can help medicine to detect and prevent diseases. Aclima is a platform created in partnership with Google, the Environmental Defense Fund, and researchers from the University of Texas at Austin, to measure air quality in big cities. A series of factors such as transport, energy consumption, and weather are taken into account. The collected data can thus be used in the prevention and management of cardiovascular and respiratory diseases [20].

2.1.2 Remote temperature store monitoring for vaccines

In less developed countries, IoMT tries to solve some apparently trivial problems, such as improving health conditions by creating conditions for storing and administering vaccines. This is how the ColdTrace System was developed, which provides remote temperature monitoring for refrigerators where vaccines are stored in rural clinics and health facilities. This way healthcare workers can safely administer life-saving vaccines [21].

Not only in less developed countries did this need to use IoT for monitoring the temperature in storage refrigerators appear. Even large pharmaceutical companies like Pfizer have adopted IoT in many ways. It has partnered with IBM and implemented IoT to help produce and distribute COVID-19 vaccines. Thus, Pfizer used IoT sensors to track and monitor shipments of COVID-19 vaccines and ensure safe temperatures over long distances [22].

2.1.3 Remote patient monitoring

The most common application of IoT devices for healthcare is remote patient monitoring security using sensors that can automatically collect biometric status values. Thus, the need for patients to travel to health centers or to collect the necessary parameters for correct monitoring was eliminated [23].

Remote patient monitoring using AI takes various forms; starting from simple systems that follow the sleep and breathing patterns of a baby (https://systemone.id)

and reaching systems that are capable of providing over 10 million diagnostic results for tuberculosis, Ebola, HIV by transmitting medical diagnostic data in real time [24].

Biometric data such as oximetry or blood pressure values obtained from patients can be uploaded to platforms such as Honeywell's Genesis Touch, which aims to keep patients connected with care providers through remote locations [25].

The IoT application in monitoring vital parameters is useful in collecting data about patients and assisting them in case of accidents. The software applications used are based on algorithms that can be used to analyze the data so that a treatment can be recommended or to generate alerts. The same principle is applied in their case of continuous and automatic monitoring of glucose levels in patients. They automatically record glucose values and can alert patients when glucose levels are problematic [26].

2.1.4 Depression and mood monitoring

There are numerous challenges related to monitoring a patient with depression or a bad emotional state because they are not always aware of the state they are in. In these situations, IoT devices can gather information about patients' depression symptoms and general mood by collecting and analyzing data such as heart rate, blood pressure, or eyeball movement [26].

A new method of monitoring these states was launched by the Abilify MyCite from Otsuka. They made an aripiprazole tablet (an antipsychotic drug used to treat various mental and mood disorders) embedded with an ingestible event marker (IEM) sensor. So the sensor sends data to a mobile app that allows users to then review "data on medication intake and activity level, as well as self-reported mood and rest quality." These data can be accessed through a secure web portal, by the attending physician, or by family members and friends [27].

2.1.5 Other examples of IoT/IoMT

IoT sensors manage to achieve continuous monitoring of Parkinson's patients' symptoms, giving patients the freedom to lead their lives in their own homes.

Apart from the portable devices presented, there are also devices in the IOMT that actually provide the patient's treatment. Some examples include devices for Hand hygiene monitoring, connected inhalers that can alert patients when they leave inhalers at home, ingestible sensors that collect information from digestive and other systems in a much less invasive way, or smart contact lenses [28].

Robots used in surgery represent an important branch of IOMT because with their help surgeons can perform complex procedures, thus reducing the size of incisions and faster healing for patients [29].

As a summary of what was previously presented, at this moment, according to the specialized literature, we can classify IoT in Biomedicine Technologies as follows:

- IoT-enabled biosensors
- Wound healing monitoring systems
- IoT-based disease monitoring systems
- Internet of nano-things healthcare applications

- Wearable IoT sensors for healthcare applications
- Degradable IoT sensors for healthcare applications
- IoT in medical implant manufacturing
- IoT in rehabilitation devices
- IoT-enabled medical robotics
- IoT in genomics
- IoT devices in pharmaceutical industries

2.2 Applications

The Internet of Things (IoT) has revolutionized the way we live our lives, yet studies show medical IoT modules are still not being used to their full potential. What is known for sure at this moment is that the Internet of Things can help transform the way health systems work and the way they provide patient care [29].

IoT in medicine is in a continuous development process and today manages to solve many medical care problems involving several levels.

Facilitating hospital management is made by room control systems, equipment monitoring and fault warning, management of equipment, medicines, and consumables, personnel performance analysis, and regulation of the flow of patients.

Improving the quality of medical services by using IoT in monitoring the vital signs of patients' health in operating and postoperative wards or online diagnostics through telemedicine solutions.

Improving the quality of the doctor-patient relationship by checking health indicators during the day with fitness bracelets, glucometers, and cuffs for measuring pulse, sending automatic reminders for activities, medications, or doctor visits, and notification of changes in vital signs with data [30] (**Table 1**).

Therefore, IoMT is a valuable technology for all players active in the health field, including public hospitals, private clinics, medical professionals of various profiles, insurance companies, and, of course, patients.

2.3 Challenges

Because we are talking about a new technology, it also faces many challenges in its application in the medical field. The main challenge is data security.

Remote patient monitoring devices cannot currently secure the collection of personal medical data. Data collected by medical devices qualifies as protected health information under HIPAA and similar regulations. As a result, IoT devices could be used as gateways for data theft if they are not secured. According to the latest studies, approximately 82% of healthcare providers report that they have experienced attacks against IoT devices.

Solutions would be the development of secure IoT hardware and software systems. Another challenge given the fact that no one at the moment can ensure that IoT devices in the health field are well managed, there is no protection system in place so that these devices are not used for other purposes than those that were created.

No	Application	Description	Problem-solving
1.	Patient care	IoT manages to ease the workflow for patient care with the help of innovative technologies that enable connectivity between medical devices saving money and time by reducing unnecessary travel of medical personnel [31, 32].	Improving the quality of medical services
2.	Patient record	The medical history part is time-consuming for the doctor, so IoT can transmit critical health data through sensors so that doctors can detect vital signs of deadly diseases in real time [33].	Improving the quality of medical services
3.	Medical assistance in medication administration	IoT helps the patient follow the medication plan correctly by issuing alarms when it is time to take their medication [34].	Improving the quality of the doctor-patient relationship
4.	Real-time patient monitoring	The use of sensors and the creation of intelligent medical systems that can be connected to a smartphone application is possible thanks to IoT technology. Thus, the data are easily collected and stored in the cloud in order to be able to monitor the patient's condition in real time [32].	Improving the quality of medical services
5.	Health Data Transmission	Medical data circulates and can be easily accessed using IoT devices. Thus, there are interconnected medical devices capable of transmitting a large amount of data in real-time applications, with IoT being responsible for constant data connectivity [35, 36].	Improving the quality of medical services
6.	Preventing and reducing the rate of intra-hospital infection	Environmental monitoring systems, hygiene control as well as pharmacy inventory tracking with the help of IoT can significantly reduce the distribution of infection among patients as well as in the hospital environment [37].	Improving the quality of medical services
7.	Making the work of doctors more efficient	IoT systems help the doctor in the relationship with the patient, relieving him of repetitive work and leaving him more time for the patient. Thus there are systems that record information about the patient with the help of voice commands and ease the doctor's work [38].	Improving the quality of medical services
8.	Telemedicine	Remote care and real-time monitoring are possible today due to the integration of IoT in medicine and the creation of a new field—telemedicine. Thus, doctors and nurses in hospitals are relieved of a large number of patients who can be monitored and cared for at home [39].	Improving the quality of medical services Improving the quality of the doctor-patient relationship
9.	IoT Hospital Management Systems	IoT-based hospital information and management systems are designed to remotely manage medical staff, medical supplies, and patient activities in the hospital. Medical staff analyze the data, which is then interpreted by hospital information and management systems. These systems are centered on the patient to increase the quality of the medical act [40].	Facilitating hospital management
10.	Rapid diagnosis	IoT systems used in monitoring can issue alerts when significant changes occur in patients' vital parameters. In this way, patients in real need of assistance can be easily identified and care teams can be directed. Thus, there is a simplification of medical procedures useful for the patient, doctors, and medical care staff [41].	Improving the quality of medical services Improving the quality of the doctor-patient relationship

No	Application	Description	Problem-solving
11.	AI and Deep Learning	Deep Learning is an AI technique used in the medical field to analyze collected data and facilitate diagnosis. The use of deep learning techniques and the mining of personal medical data collected by remote healthcare devices and sensors has revolutionized medical treatment and disease prevention [42].	Improving the quality of medical services

Table 1.
Different monitoring applications with the help of IoT.

It is also not legal to decommission patient monitoring devices that have an older version of software or firmware that makes data theft possible.

The solution to these challenges would be to correctly discover and classify all IoT devices in a healthcare provider's network. Thus, once networks of IoT devices are identified, classified, regulated, and secured in an established manner, managers can track device behavior to identify anomalies, perform risk assessments, and segment vulnerability against mission-critical devices [43, 44].

At this moment, an important challenge is represented by the final cost of medical devices. That is why studies are directed toward the design of IoT devices with sensors at affordable prices, easy to install, and maintain [45].

The general conclusions regarding the implementation of IoT in healthcare will be based on the existence of a clear and robust code of practice for data management, privacy, and cyber security. At this moment, IoT in medicine is limited to applications in the form of research projects in the field of health. The results of the current studies provide an excellent opportunity for healthcare systems to proactively predict health problems and diagnose, treat, and monitor patients both in and out of the hospital.

It is predicted that in the future, more traditional healthcare delivery practices will be supplemented or replaced by the IoT, as the adoption of technology-enabled healthcare services increases and enables healthcare systems to offer flexible models of care. In the context of IoT-enabled medical service delivery, more research is needed on the efficiency of blockchain storage compared to centralized cloud-based storage solutions (**Figure 3**).

Clinical guidelines on digital health prescriptions and the adoption of sound policies on the remuneration of primary and secondary care services delivered through IoT also need to be legislated.

All these aspects must be supported through research, in order to finally obtain a rate of acceptability and increased digital literacy of consumers and clinicians in the context of IoT use [46].

Another important aspect is that the technical preparation necessary for the implementation of IoT by those who offer medical services was not taken into account, even though the complete digitization of the medical sector is being attempted. Also, medical clinics must solve the cyber security problem and cover the lack of an adequate infrastructure for the implementation of IoT in health systems [32].

2.4 Future directions

The Internet of Things (IoT) is a fascinating technology and the possibilities of application in the medical field are limitless.

Figure 3.
Introducing medical IoT challenges.

At this moment, future research directions focus on the development of ingestible sensors and nanotechnologies that can help collect medical data in real time.

Robotic surgery has already been used in specific healthcare applications that require stable and long operational procedures.

Another field in which IoT will find applicability is the field of health insurers who will use IoT devices to calculate risk premiums with a long-term effect on patients with chronic diseases.

Companies like Google have already filed patents for contact lenses and other healthcare IoT technologies.

Health systems can be improved with the help of IoT, which can bring many benefits: simplifying decisions, reducing costs, creating better and personalized treatment plans, more efficient results, and, finally, a healthier life.

These benefits will also come with challenges such as building secure and easy-to-use IoT devices with the right software and a secure system in terms of data security [32].

3. IoT in biomedicine: Rehabilitation IoT

Health sciences are fields of study related to life and include several branches such as medicine, biomedicine, nursing, speech therapy, clinical criticism, pharmacy, physical health, dentistry, psychology, occupational therapy, nutrition, and physical therapy. All these fields consider the use of science, technology, engineering, or mathematics to provide medical care to human beings. Many people experience physical and/or motor limitations associated with a variety of reasons, whether due to problems at birth, work-related accidents, or restrictions caused by aging. Older people are more affected by motor impairments that make even simple daily tasks difficult, such as lifting an object without difficulty, eating alone, or even dressing. Such consequences can restrict personal activities and avoid the full participation of the elderly in the community, which has a negative effect on their daily work and life in general [47].

Currently, the Internet of Things (IoT) has a significant impact in the field of medical rehabilitation as well. By connecting devices and sensors to the Internet, IoT facilitates the collection and analysis of real-time data, thereby enabling medical staff to monitor patients more effectively and provide personalized treatments. Some examples of the importance of IoT in medical rehabilitation are:

1. *Patient monitoring:* IoT devices can be used to constantly monitor patients' vital parameters such as blood pressure, heart rate, and blood oxygen level. This data can be transmitted in real time to doctors and nurses, allowing them to track the patient's progress and intervene if changes or problems arise [48].

2. *Remote therapy:* IoT technologies can enable the delivery of remote therapy through Internet-connected devices. For example, patients can use wearables or mobile apps to guide them through rehabilitation exercises and collect data on their performance. Therapists can access this data and provide real-time feedback to adjust treatment and monitor patient progress.

3. *Assistive and assistive devices:* IoT can facilitate the use of assistive and assistive devices, such as smart prosthetics or robotic rehabilitation devices. These devices are equipped with sensors and can be connected to the Internet to enable data collection and real-time adjustment of operating parameters. Thus, patients can benefit from more effective and personalized rehabilitation.

4. *Inventory and equipment management systems:* In a medical environment, IoT can be used to monitor equipment and drug stocks and automatically send alerts when restocking is required. This can help to streamline processes and ensure uninterrupted operation of medical services.

5. *Data analysis and research*: By collecting and analyzing data from IoT devices, researchers and doctors can gain valuable information about the evolution of patients over time, the effectiveness of different treatments, and the factors that influence the rehabilitation process. This information can contribute to the improvement of medical practices and the development of new technologies and approaches in the field of medical rehabilitation [49].

3.1 Technologies

The architecture of an IoT (Internet of Things) system for medical rehabilitation usually involves the integration of different components and technologies to monitor, collect, and analyze data related to the rehabilitation process. A high-level architecture overview is described below:

3.1.1 Sensors and wearables

These are the physical devices that capture data about the patient's movements, vital signs, and other relevant information. Examples include motion sensors, accelerometers, heart rate monitors, electromyography (EMG) sensors, and wearable devices such as smartwatches or fitness trackers. These sensors and devices are worn by the patient and communicate wirelessly with the IoT system [50, 51].

3.1.2 Data acquisition and communication

Data collected by sensors and wearable devices is transmitted to a gateway device or directly to the cloud for processing and analysis. This communication can be done using different wireless protocols such as Wi-Fi, Bluetooth, or Zigbee. The gateway device acts as a bridge between the sensors and the cloud, relaying the data securely.

3.1.3 Cloud platform

The cloud platform serves as a central hub for data storage, processing, and analysis. It receives data from sensors and wearable devices and stores it in a secure and scalable way. The cloud platform also provides the computing resources required for real time or batch processing of the data, depending on the requirements. Popular cloud platforms such as Amazon Web Services (AWS) or Microsoft Azure are often used to host IoT infrastructure.

3.1.4 Data processing and analysis

Collected data is processed and analyzed to obtain meaningful information about the patient's rehabilitation progress. This can involve various techniques, such as signal processing, machine learning, or statistical analysis. The data processed may include metrics such as range of motion, muscle activity, exercise adherence, or performance indicators. This information can be used to tailor rehabilitation programs, track progress, and provide feedback to the patient and healthcare professionals.

3.1.5 Application and user interface

Processed data are made available to various stakeholders, including patients, caregivers, and health professionals, through user-friendly applications and interfaces. These interfaces can be web portals, mobile apps, or specialized software used in rehabilitation clinics. They provide real-time feedback, visualizations, and progress reports to help monitor and manage your rehabilitation process.

3.1.6 Security and privacy

Given the sensitive nature of medical data, security and privacy measures are crucial in a medical rehabilitation IoT system. This includes encryption of data during transmission and storage, access control mechanisms, user authentication, and compliance with relevant data protection regulations such as the United States' Health Insurance Portability and Accountability Act (HIPAA). IoT architecture should ensure that patient data is handled securely and that the system is resilient to potential cyber security threats (**Figure 4**) [52].

It is important to note that the specific architecture may vary depending on the requirements of the rehabilitation program, the type of medical condition being treated, and the available technologies. The architecture presented above provides a general framework for understanding the key components involved in an IoT system for medical rehabilitation.

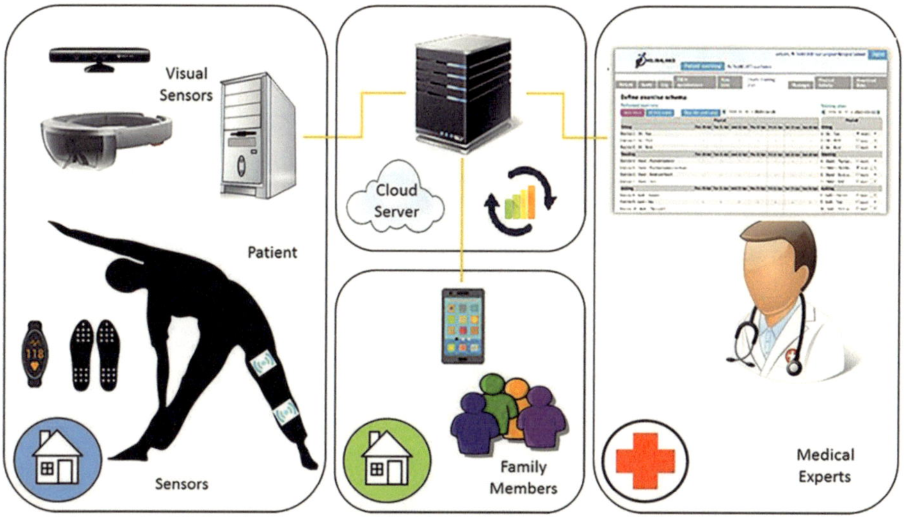

Figure 4.
System overview of the home-based rehabilitation system.

3.2 Applications

The use of IoT (Internet of Things) in medical rehabilitation can improve the effectiveness, efficiency, and convenience of rehabilitation programs. IoT data can be used to tailor rehab programs to the specific needs of individual patients. Data collected from wearables and sensors can be used to identify areas of strength and weakness, and rehabilitation programs can be designed to target these areas. Personalized approach can help patients achieve better results and reduce the risk of injury.

An example of a BSNCare+ system is where IoT-based rehabilitation equipment is embedded for both the patient and the involved environment as an end-to-end device. The nurse can use the mobile gateway to collect data in real time and offer better-quality medical services to the patient. All detection data will be transmitted to the BSN-Care server and maintained for the purpose of further data analysis and analysis of patient needs [53] (**Figure 5**).

Body Sensor Networks (BSN) take the concept of wearables to the next level. BSNs consist of a network of wearable sensors that can communicate with each other and

Figure 5.
Proposed IoT-based communication architecture for BSNCare+.

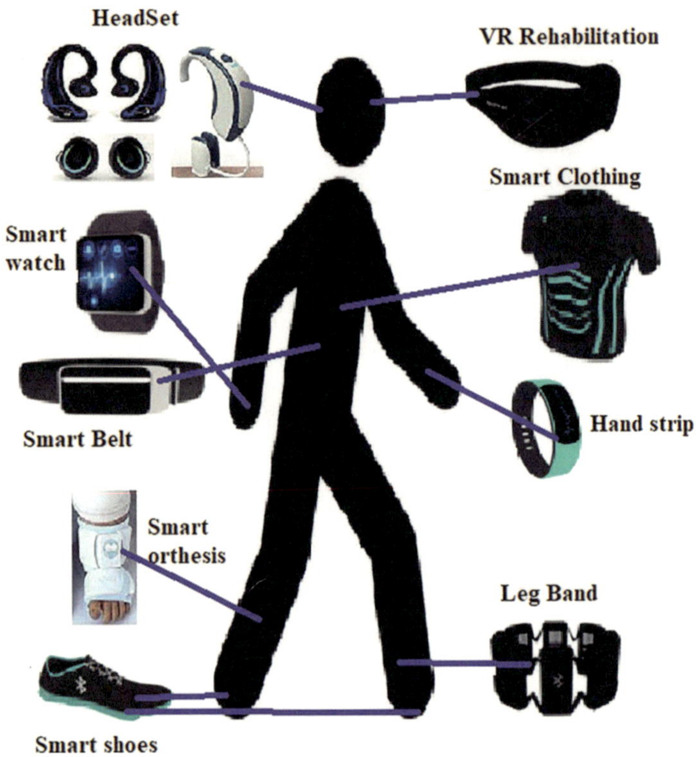

Figure 6.
Portable medical and healthcare devices worn on body parts used in medical rehabilitation IoT systems.

with other devices. BSNs can provide real-time monitoring of multiple physiological parameters, making them useful for a wide range of applications, including medical rehabilitation [54] (**Figure 6**).

IoT devices can be used to create virtual rehabilitation environments simulating real-world activities. Virtual reality headsets, for example, can be used to create immersive experiences that can help with the rehabilitation process. Virtual rehab can also allow patients to practice their rehab exercises in a safe and controlled environment.

Gamification is using game design principles to motivate and engage patients in their rehabilitation program. IoT devices can be used to create gamified experiences that make rehab more fun and engaging. For example, sensors can be used to track a patient's movements during an exercise, and the data can be used to control a video game. This can make rehabilitation more enjoyable and increase patient adherence (**Figure 7**).

IoT devices can be used to provide remote guidance and support to patients during their rehab program. Video conferencing tools, for example, can be used to connect patients with rehabilitation professionals who can provide guidance and feedback on their exercise. This can be particularly useful for patients who live in remote areas or have limited access to health professionals [55].

Overall, the application of IoT in medical rehabilitation has the potential to transform the way patients receive rehabilitation services. By providing real-time data, personalized programs, and remote coaching, IoT can improve patient outcomes, reduce costs, and increase patient satisfaction.

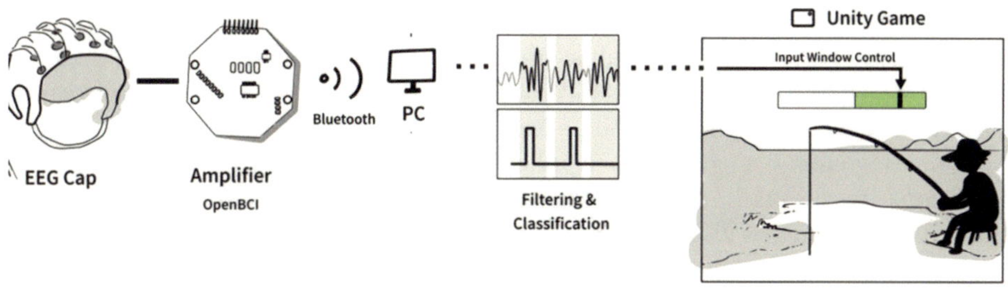

Figure 7.
Brain-computer interfaces (BCIs) and IoT for stroke rehabilitation using gamification.

3.3 Challenges

While the use of IoT (Internet of Things) in medical rehabilitation brings many benefits, there are also some challenges that need to be addressed. The accuracy and reliability of data collected by IoT devices are crucial for effective decision-making in rehabilitation. However, there may be situations where IoT devices produce inaccurate or inconsistent data due to device limitations, calibration issues, or user error. Adequate calibration, regular maintenance, and validation processes should be implemented to ensure the accuracy and reliability of IoT-generated data.

Privacy is provided by standardized data security protocols (such as encryption, authentication, and key distribution) during the data collection phase. In cases where security protocols are not suitable for IoT devices or cannot meet the application requirements, specific encryption, authentication, or key distribution algorithms are proposed. In these situations, there may also be issues related to indirect privacy leaks, such as finding user behavior patterns and protecting user anonymity. In the data transmission and storage/sharing phases, the individual system design must be made according to specific policies or rules that provide privacy to system users [56].

IoT devices and associated applications should be user-friendly and intuitive to ensure compliance and patient engagement. Complex user interfaces or difficult-to-use devices may deter patients from actively participating in their rehabilitation programs. Designing devices and interfaces with the end user in mind, conducting usability tests, and collecting user feedback can help improve the overall user experience.

IoT devices collect and transmit sensitive patient data, including personal health information. Ensuring the security and confidentiality of this data is crucial to protecting patient privacy. Healthcare providers must implement robust security measures such as encryption, secure data storage, access control, and regular system updates to protect patient data from unauthorized access or cyber threats [57]. The solution is to develop uniform data quality standards. In this way, it can be managed more efficiently across countries, organizations, and departments, thereby facilitating the storage, delivery, and sharing of data and reducing errors in judgment and decision-making due to data incompatibility, data redundancy, and data deficiencies. Since IoT systems are distributed in nature, the use of international standards can have a positive effect on improving the performance of business processes by aligning different organizations with the same foundation, addressing interoperability issues, and ultimately working in a seamless manner [58].

IoT devices in medical rehabilitation collect extensive data about patients' movements, health conditions, and activities. Healthcare providers must address ethical concerns about data ownership, consent, and use. Patients must be informed about the data collected, how it will be used, and have control over their data. Implementing clear data governance policies and complying with relevant privacy regulations are essential. Implementing IoT in medical rehabilitation may require significant investments in hardware, software, infrastructure, and ongoing maintenance. Healthcare providers need to assess the cost implications and ensure the availability of necessary infrastructure and resources to support IoT implementation effectively [59].

Addressing these challenges requires collaboration among healthcare providers, technology developers, regulatory bodies, and other stakeholders. By addressing security concerns, promoting interoperability, ensuring data accuracy, and prioritizing user experience, IoT can realize its potential to transform medical rehabilitation while providing safe and effective patient care.

3.4 Future directions

The future direction of IoT (Internet of Things) in rehabilitation is poised to bring significant advances and benefits to the field.

Here are some possible evolutions and trends:

Smart devices and sensor technology: IoT-enabled smart devices and sensor devices will continue to play a crucial role in rehabilitation. These devices can monitor and collect real-time data about patients' movements, muscle activity, heart rate, and other vital signs. This data can be analyzed to provide customized feedback, track progress, and optimize rehabilitation programs.

Adaptive and personalized rehabilitation: The IoT can contribute to adaptive and personalized rehabilitation programs. By integrating sensors into rehabilitation equipment, IoT systems can automatically adjust resistance, range of motion, or intensity based on the patient's capabilities and progress. This personalized approach can optimize the efficacy of therapy and improve patient outcomes.

Data analytics and machine learning: The large amount of data generated by IoT devices in rehabilitation presents opportunities for data analytics and machine learning. By analyzing large datasets, patterns and insights can be identified to improve rehabilitation protocols, predict patient outcomes, and optimize treatment plans. Machine learning algorithms can also help automate rehabilitation progress assessment and provide personalized recommendations.

Gamification and virtual reality: IoT can use gamification and virtual reality (VR) technologies to make rehabilitation engaging and motivating. IoT-enabled devices can connect to VR platforms, allowing patients to interact with immersive environments that simulate real-life scenarios. Gamified experiences can increase patient participation, therapy adherence, and overall motivation, leading to improved rehabilitation outcomes.

Collaborative ecosystems: The future of IoT in rehabilitation will involve developing collaborative ecosystems. Different stakeholders, including healthcare providers, device manufacturers, software developers, and researchers, will work together to create integrated solutions. These ecosystems will promote interoperability between devices, secure data sharing, and standardized protocols, encouraging innovation and progress in rehabilitation practices.

Ethical and security considerations: As IoT becomes more widespread in rehabilitation, it is critical to address ethical and security concerns. Patient confidentiality, data security, and informed consent must take precedence. Robust security measures, such as encryption and authentication protocols, should be implemented to protect sensitive patient information from unauthorized access [60, 61].

Overall, the future of IoT in rehabilitation has a great potential to transform the way rehabilitation is carried out. Using IoT technologies, healthcare professionals can provide more personalized, efficient, and accessible rehabilitation services to improve patient outcomes and quality of life.

4. Conclusions

In the medical field, IoT enables remote patient monitoring, which improves the ability to diagnose and treat various conditions. Wearable devices equipped with sensors can collect real-time data on vital signs, medication adherence, and physical activity, providing healthcare providers with valuable information about patient health. This data can be analyzed to detect early warning signs, prevent complications, and facilitate timely interventions. IoT also enables telemedicine and virtual consultations, allowing healthcare professionals to connect remotely with patients, provide guidance, and monitor their progress.

In addition, IoT plays a crucial role in rehabilitation. Connected devices can help physical therapy by tracking movements, providing feedback, and guiding patients through exercises. This real-time monitoring helps to ensure correct form and adherence to prescribed regimens. IoT-based rehabilitation tools also enable remote rehabilitation, allowing patients to receive therapy from the comfort of their homes, thus increasing accessibility and convenience. Not only does this save time and cost, it also promotes patient engagement and adherence to treatment plans.

Despite these challenges, integrating IoT into medicine and rehabilitation holds immense promise for transforming healthcare delivery. It improves patients, improves outcomes, and revolutionizes how healthcare is practiced. As technology continues to advance and healthcare systems embrace IoT solutions, we can expect to see new advancements and innovations in this area, ultimately leading to a more connected and patient-centric healthcare ecosystem.

Conflict of interest

The authors declare no conflict of interest.

Author details

Robert Fuior[1,2], Alexandru Sălceanu[1], Cătălina Luca[2] and Călin Corciovă[2*]

1 Faculty of Electrical, Energy and Applied Informatics Engineering, "Gheorghe Asachi" Technical University of Iasi, Romania

2 Faculty of Medical Bioengineering, "Grigore T. Popa" University of Medicine and Pharmacy of Iasi, Romania

*Address all correspondence to: calin.corciova@bioinginerie.ro

References

[1] Al-Fuqaha A, Guizani M, Mohammadi M, et al. Internet of things: A survey on enabling technologies, protocols, and applications. IEEE Communication Surveys and Tutorials. 2015;**17**(4):2347-2376

[2] Gubbi J, Buyya R, Marusic S, et al. Internet of things (IoT): A vision, architectural elements, and future directions. Future Generation Computer Systems. 2013;**29**(7):1645-1660

[3] Holdowsky J, Mahto M, Raynor ME, et al. Inside the Internet of Things (IoT). Deloitte University Press; 2015

[4] Parnian AR, Parsaei MR, Javidan R, et al. Smart objects presence facilitation in the internet of things. International Journal of Computer Applications. 2017;**168**(4):25-31

[5] Weyrich M, Ebert C. Reference architectures for the internet of things. IEEE Software. 2016;**1**:112-116

[6] Atzori L, Iera A, Morabito G. Understanding the internet of things: Definition, potentials, and societal role of a fast evolving paradigm. Ad Hoc Networks. 2017;**56**:122-140

[7] Jha AV, Appasani B, Ghazali AN. Performance evaluation of network layer routing protocols on wireless sensor networks. In: Proceedings of the 2019 International Conference on Communication and Electronics Systems (ICCES), July 17-19, 2019. Coimbatore, India; 2019. pp. 1862-1865

[8] Tiwary A, Mahato M, Chidar A, Chandrol MK, Shrivastava M, Tripathi M. Internet of things (IoT): Research, architectures and applications. International Journal of Future Revolution Computer Science and Communication Engineering. 2018;**4**:23-27

[9] González-Zamar MD, Abad-Segura E, Vázquez-Cano E, López-Meneses E. IoT technology applications-based smart cities: Research analysis. Electronics. 2020;**9**:1246

[10] Internet of Things in Healthcare: Applications, benefits, and challenges. Available from: https://www.peerbits.com/blog/internet-of-things-healthcare-applications-benefits-and-challenges.html [Accessed: March 6, 2023]

[11] Cvar N, Trilar J, Kos A, Volk M, Stojmenova Duh E. The use of IoT technology in smart cities and smart villages: Similarities, differences, and future prospects. Sensors. 2020;**20**:3897

[12] Chen L, Thombre S, Järvinen K, Lohan ES, Alén-Savikko A, Leppäkoski H, et al. Robustness, security and privacy in location-based services for future IoT: A survey. IEEE Access. 2017;**5**:8956-8977

[13] Shin H, Lee HK, Cha HY, Heo SW, Kim H. IoT security issues and light weight block cipher. In: Proceedings of the International Conference on Artificial Intelligence in Information and Communication (ICAIIC), February 11-13, 2019. Okinawa, Japan; 2019. pp. 381-384

[14] Gamundani AM. An impact review on internet of things attacks. In: Proceedings of the International Conference on Emerging Trends in Networks and Computer Communications (ETNCC), May 17-20, 2015. Windhoek, Namibia; 2015. pp. 114-118

[15] Javaid M, Khan IH. Internet of Things (IoT) enabled healthcare helps to take the challenges of COVID-19 Pandemic. Journal of Oral Biology Craniofactory Research. 2021;**11**(2): 209-214. DOI: 10.1016/j.jobcr.2021.01.015

[16] IoT in Healthcare: How is IoT used in MedTech? Available from: https://www.koombea.com/blog/iot-in-healthcare/ [Accessed: March 2, 2023]

[17] IoT in Healthcare: 16 Examples of Internet of Things Healthcare Device and Technology. Available from: https://builtin.com/internet-things/iot-in-healthcare [Accessed: March 2, 2023]

[18] LabOps platform [Internet]. 2020. Available from: https://elementalmachines.com/ [Accessed: March 4, 2023]

[19] IGT Systems and devices Philips [Internet]. 2022. Available from: https://www.usa.philips.com/healthcare [Accessed: March 4, 2023]

[20] Measure air. Drive solution [Internet]. 2022. Available from: https://www.aclima.io/ [Accessed: March 7, 2023]

[21] Medical Equipment Program [Internet]. 2020. Available from: https://nexleaf.org/ [March 9, 2023]

[22] Internet of things devices [Internet]. 2022. Available from: https://builtin.com/internet-things [Accessed: March 14, 2023]

[23] Ordr platform [Internet]. 2019. Available from: https://ordr.net/article/iot-healthcare-examples/ [Accessed: March 14, 2023]

[24] Miku Smart Baby Monitor [Internet]. 2021. Available from: https://mikucare.com/ [Accessed: March 20, 2023]

[25] Innovative solution [Internet]. 2020. Available from: https://www.honeywell.com/us/en#healthcare [Accessed: March 3, 2023]

[26] Ordr Clinical Defender [Internet]. 2022. Available from: https://ordr.net/article/iot-healthcare-examples/ [Accessed: March 24, 2023]

[27] Products and Medical Information. [Internet]. 2022. Available from: https://www.otsuka-us.com/ [Accessed: March 3, 2023]

[28] Technologies and services [Internet]. 2022. Available from: https://ordr.net/article/iot-healthcare-examples/ [Accessed: March 3, 2023]

[29] Technologies and services [Internet]. 2022. Available from: https://ordr.net/article/iot-healthcare-examples/ [Accessed: March 4, 2023]

[30] Software Development [Internet]. 2020. Available from: https://intellisoft.io/healthcare-software-development-services/ [Accessed: April 6, 2023]

[31] Alshammari H, El-Ghany SA, Shehab A. Big IoT healthcare data analytics ramework based on fog and cloud computing. Journal of Information Processing Systems. 2020;**16**(6)

[32] Javaid M, Haleem A, Singh RP, Rab S, Haq MIU, Raina A. Internet of things in the global healthcare sector: Significance, applications, and barriers. International Journal of Intelligent Networks. 2022;**3**:165-175. DOI: 10.1016/j.ijin.2022.10.002.

[33] Princi E, Kramer NC. Out of control–privacy calculus and the effect of perceived control and moral considerations on the usage of IoT healthcare devices. Frontiers in

Psychology. 2020;**11**. DOI: 10.3389/fpsyg.2020.582054. ISSN: 1664-1078

[34] Hao P, Wang X. A PHY-aided secure IoT healthcare system with collaboration of social networks. In: IEEE 86th Vehicular Technology Conference (VTCFall). 2017. pp. 1-6

[35] Mutlag AA, Abd Ghani MK, Arunkumar NA, Mohammed MA, Mohd O. Enabling technologies for fog computing in healthcare IoT systems, future Generat. Computing Systems. 2019;**90**:62-78

[36] Heo SP, Noh DH, Moon CB, Kim DS. Trend of IoT-based healthcare service. IEMEK Journal of Embedded Systems and Applications. 2015;**10**(4):221-231

[37] Kotha MM. Tech care: An efficient healthcare system using IoT, in. In: Emerging Research in Data Engineering Systems and Computer Communications. Singapore: Springer; 2020. pp. 655-667

[38] Kumar PM, Lokesh S, Varatharajan R, Babu GC, Parthasarathy P. Cloud and IoT based disease prediction and diagnosis system for healthcare using Fuzzy neural classifier. Future Generating Computing Systems. 2018;**86**:527-534

[39] Iqbal N, Ahmad S, Kim DH. Health monitoring system for elderly patients using intelligent task mapping mechanism in closed loop healthcare environment. Symmetry. 2021;**13**(2):357

[40] Kaur H, Atif M, Chauhan R. An internet of healthcare things (IoHT)-based healthcare monitoring system. In: Advances in Intelligent Computing and Communication. Singapore: Springer; 2020. pp. 475-482

[41] Luo E, Bhuiyan MZ, Wang G, Rahman MA, Wu J, Atiquzzaman M. Privacyprotector: Privacy-protected patient data collection in IoT-based healthcare systems. IEEE Communications Magazine. 2018;**56**(2):163-168

[42] Available from: https://www.koombea.com/blog/iot-in-healthcare/ [Accessed: April 08, 2023]

[43] Available from: https://ordr.net/article/iot-healthcare-examples [Accessed: April 8, 2023]

[44] Available from: https://intellisoft.io/the-internet-of-things-iot-in-healthcare-medicine/ [Accessed: April 12, 2023]

[45] Al-Rawashdeh M, Keikhosrokiani P, Belaton B, Alawida M, Zwiri A. IoT adoption and application for smart healthcare: A systematic review. Sensors (Basel). 2022;**22**(14):5377. DOI: 10.3390/s22145377

[46] Kelly JT, Campbell KL, Gong E. Scuffham P the internet of things: Impact and implications for health care delivery. Journal of Medical Internet Research. 10 Nov 2020;**22**(11):e20135. DOI: 10.2196/20135. PMID: 33170132; PMCID: PMC7685921

[47] Forecast unit shipments of wrist-worn wearables worldwide from 2019 to 2024. [Online]. Available from: https://healthtechinsider.com/2020/10/26/idcforecasts-double-digit-weables-growththrough-2024

[48] Dian FJ, Vahidnia R, Rahmati A. Wearables and the internet of things (iot), applications, opportunities, and challenges: A survey. IEEE Access. 2020;**8**:200-211

[49] Lounis K, Zulkernine M. Attacks and defenses in short-range wireless technologies for IoT. IEEE Access. 2020;**8**:88892-88932

[50] Gravina R, Fortino G. Wearable body sensor networks: State-of-the-art and research directions. IEEE Sensors Journal. 2020;**21**(11):12511-12522

[51] Brida P, Krejcar O, Selamat A, Kertesz A. Smart sensor technologies for IoT. Sensors. 2021;**21**:5890. DOI: 10.3390/s21175890

[52] Balakrishnan A, Medikonda J, Namboothiri PK. Analysis of the effect of muscle fatigue on gait characteristics using data acquired by wearable sensors. In: IEEE International Conference on Distributed Computing, VLSI, Electrical Circuits and Robotics (DISCOVER). 2020. pp. 137-140

[53] Yeh K-H. BSNCare+: A robust IoT-oriented healthcare system with non-repudiation transactions. Applied Sciences. 2016;**6**:418. DOI: 10.3390/app6120418

[54] Available from: https://www.iotcentral.io/blog/wearables-body-sensor-networks-smart-portable-devices

[55] Jochumsen M, Hougaard BI, Kristensen MS, Knoche H. Implementing performance accommodation mechanisms in online BCI for stroke rehabilitation: A study on perceived control and frustration. Sensors. 2022;**22**:9051. DOI: 10.3390/s22239051

[56] Yang G. An overview of current solutions for privacy in the internet of things. Frontier in Artificial Intelligence. 2022;**5**:812732. DOI: 10.3389/frai.2022.812732

[57] Renard Y, Lotte F, Gibert G, Congedo M, Maby E, Delannoy V, et al. Openvibe: An open-source software platform to design, test, and use brain–computer interfaces in real and virtual environments. Presence Teleoper: Virtual Environment. 2010;**19**:35-53

[58] Zhang L, Jeong D, Lee S. Data quality management in the Internet of things. Sensors (Basel). 2021;**21**(17):5834. DOI: 10.3390/s21175834

[59] Papan J, Segec P, Yeremenko O, Bridova I, Hodon M. Enhanced multicast repair fast reroute mechanism for smart sensors IoT and network infrastructure. Sensors. 2020;**20**:3428

[60] Zikria YB, Ali R, Afzal MK, Kim SW. Next-generation internet of things (IoT): Opportunities, challenges, and solutions. Sensors. 2021;**21**:1174. DOI: 10.3390/s21041174

[61] Atzori L, Iera A, Morabito G. The internet of things: A survey. Computer Networks. 2010;**54**(15):2787-2805

Chapter 5

A Survey on IoT Fog Resource Monetization and Deployment Models

Cajetan M. Akujuobi and Faith Nwokoma

Abstract

There has been an immense growth in the number of applications of devices using the Internet of Things (IoT). Fog nodes (FN) are used between IoT devices and cloud computing in fog computing (FC) architecture. Indeed, an IoT application can be fully serviced by local fog servers without propagating IoT data into the cloud core network. FC extends the cloud-computing paradigm to the network edge. This paper surveys fog resources monetization and the wide use of IoT devices in making FC a paramount technology necessary to achieve real-time computation of IoT devices. We looked into the monetization architectures applied by various literature. We found that the decentralization fog monetization architecture stands out since it solves some issues posed by centralized fog monetization architecture, such as QoS and additional fee costs by third parties payment gateway.

Keywords: Internet of Things (IoT), fog computing (FC), fog monetization architecture, blockchain, quality of service (QoS)

1. Introduction

The application of Internet-of-Things (IoT) technology to almost every sphere of life has resulted in immense growth in the number of IoT devices, the acceptance of IoT, IoT applications, and the volume of data uploaded to cloud systems. The international data corporation (IDC) forecasted that about 41 billion IoT-connected devices will be active in 2025, producing data surpassing 79 ZB [1]. Current cloud systems are not large enough to process and store this increase in IoT data traffic to meet real-time demands [2], which affects all IoT systems. Too many networks towards a distant cloud can cause high latency for sensitive IoT applications, such as healthcare [3], multimedia [4], and vehicular/drone applications [5, 6]. In addition, the centralization of the cloud may lead to a reduction of privacy in IoT data uploaded [7].

Fog nodes (FN) are used between IoT devices and cloud computing in fog computing (FC) architecture to reduce the distance data travels for processing in the cloud. Therefore, enabling cloud computing services from core network infrastructures to customer premises by using fog nodes, namely; switches, private servers, cloudlets, routers, etc. The closeness of fog nodes to edge devices results in a large reduction in latency, energy-efficient, and optimal use of the network bandwidth for applications within agriculture, smart cities, etc. [8]. Also, fog nodes eliminate data duplication and empower applications using fog with local and near real-time intelligence. Regarding processing and storage abilities, fog servers are much smaller than cloud [9]. Still, fog servers' larger number and geo-distribution allow fog to alleviate cloud network congestion by servicing many IoT applications [8]. Indeed, an IoT application can be fully serviced by local fog servers without propagating IoT data into the cloud core network. Fog computing enables computational workload offloading through fog nodes which can further reduce the transmission latency and ease traffic congestions on the Internet. It also introduces many new services and applications that cannot fit the traditional cloud computing architecture well. For example, large-scale environmental monitoring systems can deploy computationally intensive applications at the sensors and utilize the fog computing architecture to achieve instantaneous response [9, 10].

This paper presents a survey on fog resources monetization, a payment system implemented for the services delivered by fog resources [11]. The wide use of IoT devices has made FC a paramount technology necessary to achieve real-time computation of IoT devices. The basic unit of FC, which is the FN, is defined in this literature, also highlighting the characteristics of FC. The available deployment and revenue models are divided into four and discussed briefly. Fog resource monetization was divided into centralized and decentralized architectures. The centralized architecture has a central authority, which determines the pricing model and quality of service (QoS), while the decentralized system has no central authority and no fixed pricing model.

2. Background

The term "Internet of Things," composed in 1999 by Kevin Ashton [12], refers to a network of interconnected physical devices, technologies, objects, and services through the Internet for the exchange, processing, and storage of data [13, 14]. IoT devices use sensors to get information from the surrounding environment and regularly respond to this information through an actuator. The application of IoT has evolved over the years, and it is widely used to implement a variety of groundbreaking smart devices and services. Unfortunately, it cannot often implement such services directly through the IoT device. The large variety of heterogeneous data, otherwise known as big data [15], must be computed and stored by IoT devices. IoT devices are mostly battery-powered and have deficient networking, processing, and storage resources. Therefore, it cannot efficiently carry out these operations (computing, storing, and networking) [12]. The cloud-computing concept was introduced to compensate for IoT devices' deficiencies in computing, processing, and storing big data [2].

The exponentially growing request for computationally intensive applications and services makes cloud computing necessary. Cloud computing gives users access to various on-demand services by judiciously using the hardware and software in cloud data centers [12]. Large-scale data centers are huge and costly; therefore, they are always built-in low-cost remote areas. Although cloud computing solves the computational and storage problem of IoT devices, it poses a challenge of high latency due to its distance from the IoT devices located at the edge of the network and poor quality of service (QoS) due to data traffic in the network between the cloud and edge devices [10]. A new framework called fog computing was proposed to resolve the issues associated with cloud computing.

The concept of fog computing can be traced back to early 2009 when Satyanarayanan et al. [16] proposed using cloudlets to cope with the limits of cloud computing, especially the high and unpredictable latencies. These cloudlets provide the benefit of cloud computing close to the edge devices, and when there are no cloudlets, the edge devices communicate with the cloud directly. Cloudlets are used to support edge devices to carry out computational operations. This process of using cloudlets is called edge computing [17]. "Edge computing" and "fog computing" are often used interchangeably. However, there are similarities between them. It's essential to identify that edge computing does not view the overall service as consisting of a hierarchy of nodes with the cloud also included; instead, the overall service is performed by a close-by cloudlet [17]. Due to this, Open Fog Consortium differentiates fog and edge computing, highlighting that fog works with the cloud and is hierarchical. In contrast, the edge works independently of the cloud and is restricted to several layers. Hence, it is important to note that, though there are similarities between the two concepts, they were designed for different contexts; nevertheless, they are both growing towards an inevitable convergence [18–21].

FC extends the cloud-computing paradigm to the network edge. Formally, fog computing (FC) is defined as the virtualization of network architecture that "uses one or a collaborative multitude of end-user clients or near-user edge devices to carry out a substantial amount of storage (instead of stored primarily in cloud data centers), communication (instead of routed over backbone networks), control, configuration, measurement, and management" [22]. Fog computing is proposed to enable computing directly at the network's edge, which can deliver new applications and services, especially for the future of the Internet of Things. Fog computing uses fog nodes at the edge to interact with IoT devices.

3. Definition of fog node

Fog computing overcomes the limitations of cloud computing to enable real-time analysis for smart devices at the edge of the network. The process of fog computing involves data transfer to fog nodes for processing, storage (temporary storage), and networking operations from edge devices. Fog nodes are the basic units of fog computing. A network device that uses processing capabilities, dedicated servers, or computational servers to coordinate underlying edge devices can be referred to as a fog node [21, 23], or, put, a fog node is a physical device that performs fog computing. In [21], some examples of fog nodes were given as; wireless access points, routers, video surveillance cameras, switches, and Cisco Unified Computing System (UCS) servers. One uniform feature among all these devices is that they all embed storage, computing, and networking abilities, all necessary for IoT applications. A fog architecture is usually an aggregation of several levels of nodes. A processing application might be suited to a particular level due to the specifics of the requirement of that application for such features as latency, mobility, security/encryption, and the need for quick scalability [24]. The position and number of levels of fog nodes in a hierarchical fog will depend on the architecture involved. In the architecture described in [25], fog nodes are created near base stations in 5G networks. In contrast, as described in the architecture in [26], end users contribute to providing fog devices within residential areas and are rewarded as incentives to share fog nodes.

4. Fog computing characteristics

Fog Computing is a highly virtualized platform that overcomes the limitation of interaction between end devices and the cloud. Fog computing devices are mostly located at the edge of the network; they bring the paradigm of cloud computing, such as computing, networking, and storage services, to the edge of the network. It provides all the benefits of the cloud to the edge devices and compensates for the limitations of the edge and IoT devices. Fog computing enhances the performance of edge computing by reducing the time, bandwidth, and energy requirements that would have been expended in IoT-Cloud communication. In this section, the characteristics of fog computing from the works of [12, 21, 27–30] are highlighted below:

4.1 Edge location

One of the major characteristics of fog computing is that it contains fog nodes located at the edge of the network. The FNs are in the same environment and location where the IoT devices generate their data [31]. From the perspective of the Communication Service Provider, FNs are the cloudlets attached to the base stations that are distributed with the service masts/tower [31]. These fog nodes enable the computational ability of the cloud to be performed at the edge of the network.

4.1.1 Location-awareness

Fog computing supports location awareness applications. The ability of an edge device to be aware of its location through an application is known as location awareness. Location awareness enables location-specific services and information to be available for users when a device enters or leaves a geographical region. This is particularly important for mobile edge IoT devices in applications like the automotive, drone, and health industries. Location-awareness features of the FC network provide important information in resource planning and distribution for equitable, even, and fair service distribution.

4.1.2 Low latency

The issue of high latency gives rise to the need for FC due to the distance between edge devices and cloud systems. FC alleviates this high latency by providing the network's edge fog nodes. FN supports end devices with cloud services at the network's edge, including applications with low latency requirements (e.g., gaming, video streaming, and augmented reality). FNs make the cloud's robust computational and storage capacities available to the edge devices in the shortest time possible since they are located almost in the same network and environment as the IoT and other edge devices needing their services [32].

4.1.3 Geographical distribution

The cloud is more centralized, while in contrast, FC uses services and applications that requires vastly distributed deployments. FC contains varieties of FN, which are widely distributed, and are placed in different places such as highways, tracks, network infrastructure, and even in residential buildings. Therefore, fog computing consists of widely distributed fog nodes that enable data processing, storage, and computing with IoT devices.

4.2 Large-scale sensor networks

Fog computing has a very large-scale sensor network. These networks are connected between the fog nodes, clouds, and endpoint devices. Due to fog computing having a large-scale network since it is widely distributed, it can be used to monitor environments, smart city designs, smart agriculture implementations, and smart grid applications. These large-scale sensors send data continuously to the FN for processing, analysis, decision-making, and storage. These find applications in smart home devices, wearable devices, industrial sensors, connected appliances, smart healthcare devices, vehicles, environment monitoring devices, and smart agriculture devices.

4.3 Large number of fog nodes

Fog computing supports a very large number of IoT devices with cloud paradigms. Due to the large number of IoT devices that are widely distributed, FC has a large number of FNs to support these IoT devices. Fog nodes, sometimes called fog servers, include servers, routers, gateways, and IoT devices with routing, storage, and computing capabilities.

4.4 Support for mobility

It is essential for many Fog applications to communicate directly with mobile devices and therefore support mobility techniques since a good percentage of the devices on the edge of the network are not stationary, for example, wearables, drones, and self-driven cars. FC must have robust mobile support capability for efficient edge computing. Mobile devices, like automobiles and drones, always change location quickly and depend on FN's critical services for operational efficiency and decision-making. This will depend on the capability of the fog network to offer these services without a drop in the quality of services rendered as these devices move from one location to another [33].

4.5 Real-time interactions

An important feature of FC design is the need for real-time support for edge devices. FC is designed to greatly reduce the latency in communication between IoT and the cloud [34]. Fog applications involve real-time interactions rather than batch processing. Fog computing enables real-time interactions between end devices and fog nodes by ensuring it operates at the lowest possible latency. The real-time feature of the fog supports gaming, healthcare, automotive, aviation, streaming, and security systems.

4.6 Heterogeneity

IoT comprises different devices, including Fog nodes deployed in various environments from different vendors and technologies. Fog Computing, as a platform of high virtualization, yields computation, storage, and networking services, bridging the gap between edge devices and the cloud. While standardization has not been achieved across various FC computing paradigms like deployment methods, orchestration strategies, and equipment designs, there is beginning to be convergence between enterprises of similar interests [31]. Fog computing must continue to grow in accommodating heterogeneous vendors for equipment and application. It is also necessary to have communication protocols that will assist in interoperability.

4.6.1 Scalability/flexibility

Resources and devices should be added dynamically to accommodate constant changes in the network system. Fog networks should be distributed and have the flexibility of ease of integration with new devices and other networks. It should be scalable to meet the ever-growing deployment of IoT devices and to handle the enormous data generated from these IoT devices for processing, analysis, and storage. FC has the scalability features of the cloud for quick provisioning and an increase in available computing resources to handle spikes in service requests.

4.7 Interoperability and federation

For the seamless delivery of some peculiar services, like streaming, the cooperation of several fog providers will be required. For this reason, there is a requirement for fog components to interoperate, and services will need to have the capacity to be federated across different domains. Like cloud services, FC is delivered across different layers of technologies and paradigms. You have the ISPs, cloud services providers, payment industries, network equipment vendors, varying communication protocols, and security standards, to mention a few. However, to a large extent, there is a federation in most cloud paradigms enhanced by standardizations across these layers, which has enhanced interoperability, and there is still continuous to be more. For FC to meet the emerging IoT service needs, more must be done in federation across vendors and communication stack to ease the different technologies' interoperability.

4.7.1 Filtering

Edge devices produce huge amounts of data in real time, and the fog nodes filter all this data. Some filtered data are sent to the cloud for further processing or storage. The edge device that directly receives the data from the IoT devices must be able to determine noise in the data and perform some level of processing and analysis that will support the IoT's immediate need before offloading it to the cloud. This enables real-time analysis of data by the fog. It also saves the bandwidth wasted in offloading useless data to the cloud. The filtering capability of the FC allows faster data analysis and decision-making at the edge.

5. Deployment models and revenue scenarios of fog computing

Understanding the different revenues and incentive structures employed through different deployment models is a challenge necessary to enable a vast adoption of fog computing systems. The different revenues and incentive models proffer a better understanding as to why:

i. Infrastructure providers would offer their resources to act as FN

ii. Users would want to make use of these FC resources.

Edge architecture, such as Wi-Fi deployments within cities, can be considered similar to FC deployment, which different organizations control. Characteristics of FC, like geographical distribution, security requirements, and FN heterogeneity, are related to the revenue models. They play a major role in how FNs can generate a potential revenue stream for FN providers [12]. Maintaining a suitable infrastructure with good computing network performance and power without sufficient FNs will be unrealistic. Providing incentive models for the provision and maintenance of FNs is essential. We consider the following four types of deployment models. The description below attempts to provide context for the deployment model based on the particular deployment approach being used in [12, 31, 35–37].

5.1 Dynamic FN discovery supported revenue model

This model describes the dynamic discovery of an FN as an end device changes its location. The user device searches for FN in its "vicinity" using the advertised profile of the node (which can include availability statistics, security credentials, and types of available services). Applying this approach, the user is not guaranteed that a suitable FN will be found to sustain an application session. Still, negotiation can take place if multiple fog nodes are found. A user device can also cache previously seen fog nodes. The incentive for the provider is to gain revenue from each user session sustained using that FN. A user is charged based on connection time, size of data, or range of services utilized. The incentive for this deployment model is based on the fact that fog node providers gain revenue from each user session based on the connection time, range of services utilized, and size of data, therefore it is necessary for the fog nodes provider to make the FN discoverable to enable users to connect. The revenue earned by undertaking this would be the basis for the deployment model. This deployment model gives the user the option to choose the fog node needed based on the service and subscription model provided by the FN.

5.2 Pre-agreed contracts with fog providers

This deployment model generates pre-agreed contracts with operators of specific FNs—negotiated at a set price. Hence, there would be a preferential selection of particular nodes by a user if multiple choices are found. This also reduces user risks, as security credentials would be included in these pre-agreed contracts and could be configured (e.g., use of particular encryption keys) beforehand. These pre-agreed contracts must comply with service-level objectives (e.g., an availability profile) that an operator needs to meet. It is, therefore, possible that a fog node operator will outsource their task to a Cloud provider. The incentive for the provider is to increase the number of potential subscribers by developing pre-agreed contracts. Capacity planning associated with such FNs depends on accurately predicting potential future demand. In this case, the deployment model involves agreeing to a cost for entering into a contract with a Fog provider. This contract also provides preferential access to the provider's fog nodes.

5.3 FNs federation

This deployment model involves multiple FN operators collaborating to share the workload. This would imply a federation between FNs within a particular geographical area to sustain potential revenue. There would be a preferred cost for sharing the workload with other providers, enabling revenue sharing between providers. It is necessary to identify how workload "units" can be characterized to enable such an exchange. This is equivalent to alliances between airline companies, where specialist capability (and capacity) available along a particular route can be shared across multiple operators. In the same way, if an operator deploys specialist GPUs or video analytics capability within an FN at a particular location, other operators could also seamlessly make use of this and similarly share other capabilities in other locations. This type of geographic-centric specialization could enable localized investment within particular areas by operators.

5.4 Fog-cloud exchange

This deployment model involves a user device not being aware of the existence of any FN. Instead, the user device interacts with a Cloud operator who then attempts to find an FN near the user. Therefore, the Cloud operator needs to keep track of the user location and discover suitable FN operators that could be used to support the session at a particular location. In this instance, the Cloud operator will always try to complete the user request first; however, if a QoS target is unlikely to be met due to latency constraints, it can outsource the user request to a regional FN. The incentive in this instance is to enable Fog-Cloud exchange contracts to be negotiated between providers.

6. Evaluation of fog computing resources monetization architecture

Fog computing is well known for making cloud processing, storing, and computational ability available at the network edge through fog nodes. Therefore, enabling cloud abilities to be carried out from end devices to nearby fog nodes, hence high computational power and low latency are achieved simultaneously. Fog architecture performs a huge role in fog computing. In most related works, the fog computing architecture is described as the structure shown in **Figure 1** [25, 38–40].

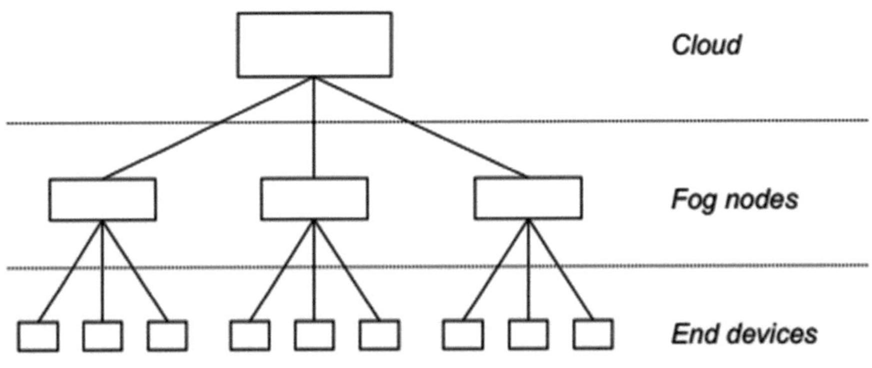

Figure 1.
Abstract model of fog computing.

This structure involves an end device communicating with the fog nodes and clouds to request or send data for processing or storage. The monetization aspect of this architecture introduces a fourth block, either a third-party operator or a smart contract. Fog node resource monetization varies according to the literature. In this work, the architecture based on the monetization and pricing model employed is divided into two, namely:

I. Centralized monetization architecture

II. Decentralized monetization architecture

7. Centralized monetization architecture

According to [35, 41, 42], the centralized monetization architecture is shown in **Figure 2**. This architecture comprises the cloud, fog nodes, edge devices, and third-party payment gateway. The third-party payment gateway is an entity that helps the fog providers receive payment online for the services rendered to the end users. This is called a centralized monetization architecture because the fog provider has firm control and authority over the kind of services rendered and determines how the fog services will be monetized, irrespective of the QoS provided. The third-party payment gateway implements the pricing and monetization strategies between edge devices and fog nodes. This leads to a subscription-based pricing model [41]. This

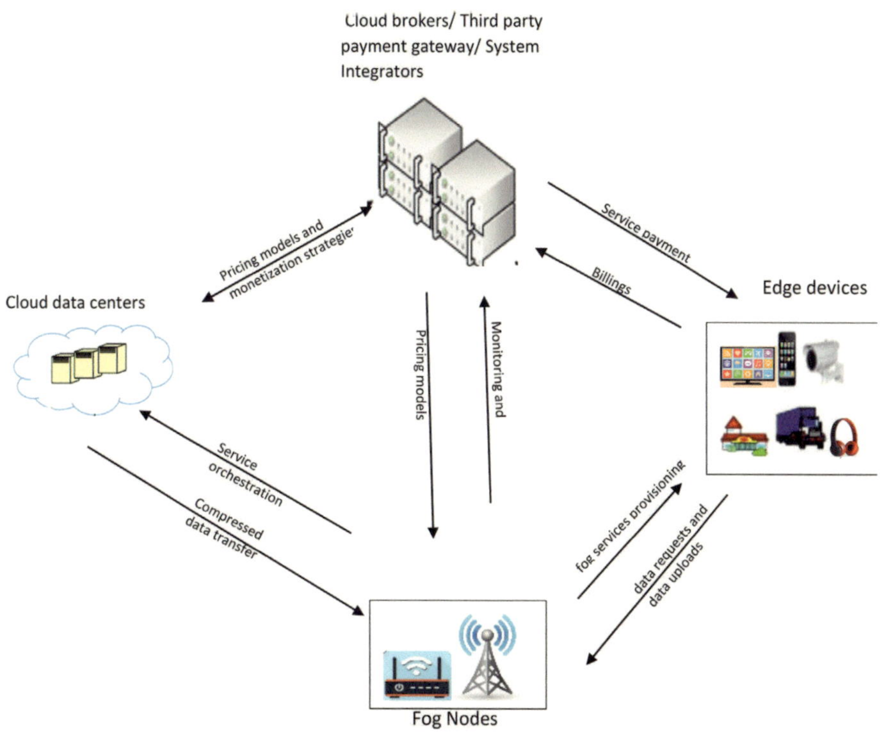

Figure 2.
Central monetization architecture of fog nodes.

fixed payment is only advantageous when the fog service providers deliver the quality of service promised. Still, there is mostly a variation in the quality of service (QoS) which is not reflected in the subscription-based pricing models. The QoS promised is not always the same in all instants when the customer accesses the fog service, resulting in mistrust between the fog node provider and customer. Also, service charges by third parties increase the cost of using the fog services by the consumers. Since it is centralized and embeds a fixed pricing model, hence once the promised quality of service is not met, it might lead to customers churn and vendors lock-in for situations where it is difficult for customers to migrate to another vendor due to sole dependence on a particular vendor [35].

8. Decentralized monetization architecture

The decentralized monetization architecture shown in **Figure 3** consists of cloud servers, public fog nodes, edge devices, and a smart contract whose major function is ensuring monetization and a structured, logical revenue exchange between fog nodes and edge devices. In this kind of monetization architecture, the cloud or fog provider has no control or authority over the monetization and pricing model of services rendered by the fog nodes. The monetization strategy is shared between the fog service provider and the end user through a smart network, for example, blockchain. The blockchain network accesses the quality of service the fog devices provide and determines the pricing model to employ between the end user and the fog service provider.

The Ethereum smart contract was the monetization smart contract [35]. The smart contract was divided into fog node provider only, device only, and fog only as the layers of authorization provided in the Ethereum smart contract, and each entity, as the name of the layer suggests, can only access layers present to it. The edge devices

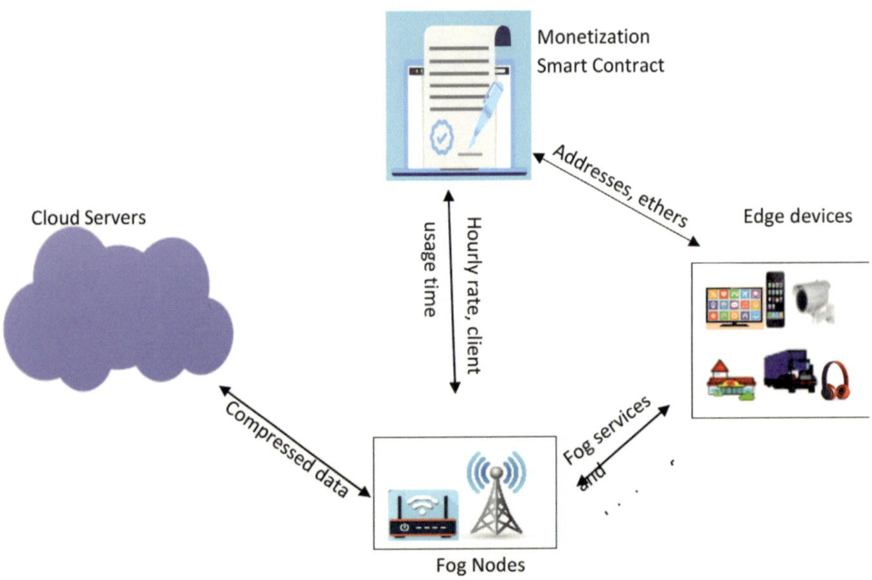

Figure 3.
Decentralized monetization architecture of fog node.

are registered in the smart contract, and the device deposits an initial amount. The interaction between the fog node, Ethereum smart contract, and edge devices enables money to be paid to the fog node for services rendered or money refunded by the fog node if there is a breach in trust between the fog nodes and the device. Also, an individual or a particular organization may own a set of fog nodes. These fog nodes generate, curate, and process raw data in a specific geographical area. These processed data may not all be needed by the organization. Lizcano [43] presents a way of monetizing these fog nodes with other fog nodes owned by an individual or organization needing those processed data. Guevara et al. [24] propose a digital marketplace where fog nodes requiring a specific data set can connect with others to get the required data using blockchain and FIWARE technology. The blockchain ensures trust between fog nodes during data exchange and implements the pricing model configured in the smart network while data is being exchanged. FIWARE technologies ensure the interoperability of data between the fog nodes. The blockchain ensures trust and nonrepudiation between the devices, while the FIWARE technologies ensure interoperability between the connected fog nodes. Furthermore, fog computing depends on joint action between several infrastructure operators and service providers who manages and operates this infrastructure pose a major challenge. Also, resource allocation is still a major problem for fog computing instances. A user-participatory fog computing architecture was proposed in [25]. This fog architecture is similar to a WIFI architecture where users connect to the WIFI with their devices. Likewise, in the model, services provided by fog benefit the users when they install fog devices to the network. In contrast, the fog container placement is controlled by fog managers to make it feasible. After successfully connecting to the network, the user registers fog devices in the fog portal. The fog portal between the corresponding resources plays an intermediary role.

The decentralized fog monetization architecture obliterates the issues of QoS and third-party fees faced by centralized architecture. The decentralized architecture employs smart contract technology with algorithms written for the monetization and pricing model. The interaction between the public fog nodes and the edge devices is made public to all network members through a public ledger. The trust between the customer and fog node providers is restored, and the quality of service is tracked at each stage. Once it drops below a certain standard, there is a breach of trust between the fog node and the edge device, and a refund of revenue is demanded [23]. Fog nodes with more trust issues are flagged by the blockchain network and avoided by other fog devices. Their system reputation is monitored, which keeps the fog providers in check.

9. Challenges

Although there is a vast improvement in ideas related to Fog node monetization, some challenges are still encountered, which need to be addressed. To make fog computing a reality, to the extent of the demands postulated, some of the open challenges of fog are listed below:

1. *Fog Networking*: The heterogeneous nature of the fog network placed at the internet edge poses a challenge in managing and controlling services such as maintaining connectivity between heterogeneous devices. Many vendors and major players are currently providing fog networks with silo technologies. This makes

interoperability difficult, and it is a challenge for the minor players to contribute or enter the fog network ecosystem. The convergence of the major providers, like Amazon, Microsoft, Ericson, etc., to a common standard, will boost the expansion of the fog market. As seen in [31], some bodies are beginning to come together to create a common ground for technology convergence. These include the 5G Alliance for Connected Industries and Automation (5G-ACIA), Automotive Edge Computing Consortium (AECC), Industrial Internet Consortium, and many others. As noted in [31], while standardization bodies like 3GPP, ETSI, and TM Forum and pushing out standards for fog and IoT, some open source forums are also contributing to fog convergence, such as Cloud Native Computing Foundation (CNCF), Open Network Automation Platform (ONAP) and LF Edge [29]. More research must be conducted to show ways of providing these services more flexibly.

2. *Task Scheduling*: Task scheduling is not an easy fix in the Fog. This is because the task can move between various physical devices like fog nodes, back-end cloud servers, and client devices. There is a need for efficient communication between different planes, administrative, data, user interface, and many other processes for seamless task scheduling. Caprolu et al. [44] explored similarities with docker for containerization. Since many fog technologies, especially open-source forums [31], are moving towards an autonomous distributed system, more work must be done in designing an efficient system that will distribute service requests among the many serving nodes considering the features of such services. For instance, a pool of IoT devices seeking data offload and analysis much be merged with a fog node with compute and storage capacity, and the location of the serving nodes must also be accounted for concerning latency requirements.

3. *Management*: For fog computing to be feasible, there are potentially billions of small edge devices and fog nodes to be configured, the fog will heavily rely on decentralized (scalable) management mechanisms that are yet to be tested, and this, at an unprecedented scale. The technology can only deliver its best decentralized. Like blockchain, FC needs to support technologies like smart cities, smart agriculture, automotive, and surveillance systems without the control of a central firm. The management plan must be flexible so that new players will not find it difficult to enter the market. Service requests and integrations must be made seamless for end users. Such an ecosystem will lead to a rise in fog technologies and service quality. Some of these issues are partly laid out in [45].

4. *Location of Fog Nodes*: Fog nodes are the basic unit of fog computing. As the number of IoT devices increases exponentially, more fog nodes will also be needed. The problem becomes; where will these fog nodes be placed to ensure optimal functionality? Many providers of fog nodes leverage telecommunication site locations for fog servers. This is not efficient as telecom sites are not evenly distributed. This deployment type will disfranchise industries like agriculture, usually in rural areas. Fog technology integration technologies need to make room for user-contributed nodes so individual users can contribute FN to locations with a deficiency of FN. Incentives for node contributions must be lucrative enough to attract sufficient contributions.

10. Conclusion

Fog computing has immensely alleviated the challenges faced by edge computing, and has introduced us to new possibilities we can achieve with real-time analysis. The application of fog computing is vast, ranging from health care, agriculture, sports, housing, computations, etc. Fog computing makes up all the deficiencies of IoT networks in computing and storage and reduces the latency in IoT-Cloud communication. Monetization of fog computing has been a major challenge since all participants seem not to be favored in any system of monetization. In this work, we surveyed the characteristics of fog computing, such as a large number of nodes, edge location, low latency, etc. We briefly explained the working relationships, the importance of each, and how fog computing brings cloud capabilities to the edge. We also explored the monetization architectures applied by various literature. We found in the study that the decentralization of the fog monetization architecture stands out since it solves some of the issues posed by centralized fog monetization architecture, such as QoS and additional fee costs by third parties payment gateway.

Author details

Cajetan M. Akujuobi* and Faith Nwokoma
Center of Excellence for Communication Systems Technology Research (CECSTR), Electrical and Computer Engineering, Prairie View A&M University, Prairie View, Texas, USA

*Address all correspondence to: cmakujuobi@pvamu.edu

References

[1] Wright A. The growth in connected IoT devices is expected to generate 79.4ZB of data in 2025, according to a new IDC forecast. Framingham, MA, USA, Rep. PrUS45213219: International Data Corporation (IDC); 2019. Accessed: April 2020. [Online]. Available from: https://www.idc.com/getdoc.jsp?containerId=prUS45213219

[2] Fog Computing and the Internet of Things: Extend the Cloud to Where the Things Are. San Jose, CA, USA, Rep. C11-734435-00: CISCO; 2015. Accessed: April 2019. [Online]. Available: https://www.cisco.com/c/dam/en_us/solutions/trends/iot/docs/computing-overview.pdf

[3] Santos GL et al. Analyzing the availability and performance of an e-health system integrated with edge, fog and cloud infrastructures. Journal of Cloud Computing. 2018;7(1):16

[4] Do CT, Tran NH, Pham C, Alam MGR, Son JH, Hong CS. A proximal algorithm for joint resource allocation and minimizing carbon footprint in geo-distributed fog computing. In: IEEE Proceedings of the 2016 International Conference on Information Networking (ICOIN). 2015. pp. 324-329. [Online] Available from: https://ieeexplore.ieee.org/document/7057905

[5] Yu C, Lin B, Guo P, Zhang W, Li S, He R. Deployment and dimensioning of fog computing-based internet of vehicle infrastructure for autonomous driving. IEEE Internet Things Journal. 2019;**6**(1):149-160

[6] Loke SW. The internet of flying-things: opportunities and challenges with airborne fog computing and mobile cloud in the clouds. arXiv:1507.04492. 2015. [Online]. Available from: https://arxiv.org/abs/1507.04492

[7] Vaquero LM, Rodero-Merino L. Finding your way in the fog: Towards a comprehensive definition of fog computing. ACM SIGCOMM Computer Communication Review. 2014;**44**(5):27-32

[8] Bonomi F, Milito R, Zhu J, Addedapalli S. Fog computing and its role in the internet of things. In: Proceedings ACM 1st Ed. MCC Workshop Mobile Cloud Computing. ACM Digital Library; 2012. pp. 13-16. DOI: 10.1145/2342509.2342513. Available from: https://dl.acm.org/doi/10.1145/2342509.2342513

[9] Vahid Dastjerdi A, Gupta H, Calheiros RN, Ghosh SK, Buyya R. Fog computing: Principals, architectures, and applications. ArXiv: 1601.02752. 2016. Available from: https://arxiv.org/abs/1601.02752

[10] Yi S, Hao Z, Qin Z, and Li Q. Fog Computing: Platform and Applications

[11] Farshad F, Bahar F, Mahmoud D, Cesare P. Guest editorial: Special issue on AI-driven IoT data monetization: A transition from Value Islands to value ecosystems. IEEE Internet of Things Journal. 2022;**9**(8):5578

[12] Bassi A, Bauer M, Fiedler M, Kramp T, Kranenburg RV, Lange S, et al. Enabling Things to Talk. NY, USA: Springer Heidelberg, Springer Link; 2013. DOI: 10.1007/978-3-642-40403

[13] Ray PP. A survey on internet of things architectures. Journal of King Saud University-Computer and Information Sciences. 2016;**30**(3):291-319

[14] Ni J, Zhang K, Lin X, Shen XS. Securing fog computing for internet

of things applications: Challenges and solutions. IEEE Communications and Surveys Tutorials. 2018;**20**(1):601-628

[15] Zhou Z, Yu H, Xu C, Chang Z, Mumtaz S, Rodriguez J. BEGIN: Big data enabled energy-efficient vehicular edge computing. IEEE Communications Magazine. 2018;**12**:82-88

[16] Satyanarayanan M, Bahl P, Caceres R, Davies N. The case for VM-based cloudlets in mobile computing. IEEE Pervasive Computing. 2009;**8**(4):14-23

[17] Ha K and Satyanarayanan M. OpenStack++ for Cloudlet Deployment. Technical Report. CMU School of Computer Science. 2015. Available from: http://elijah.cs.cmu.edu/DOCS/CMU-CS-15-123.pdf

[18] Satyanarayanan M. The emergence of edge computing. Computer. 2017, 2017;**50**(1):30-39

[19] Satyanarayanan M, Chen Z, Ha K, Hu W, Richter W, Pillai P. Cloudlets: At the leading edge of mobile cloud convergence. In: Proceedings of the 6th International Conference on Mobile Computing, Applications, and Services (MobiCASE'14). Ericsson; 2014. pp. 1-9

[20] Satyanarayanan M, Simoens P, Xiao Y, Pillai P, Chen Z, Ha K, et al. Edge analytics in the internet of things. IEEE Pervasive Computation. 2015;**14, 2015**(2):24-31

[21] Yi S, Li C, Li Q. A survey of fog computing: Concepts, applications and issues, proceedings of the 2015 workshop on Mobile big data. ACM, Mobidata 15; 2015:37-42

[22] Chiang M. Fog Networking: An Overview on Research Opportunities. arXiv preprint arXiv:1601.00835. [Online]. Available from: http://arxiv.org/pdf/1601.00835

[23] Marín-Tordera E, Masip-Bruin X, García-Almiñana J, Jukan A, Ren GJ, Zhu J. Do we all really know what a fog node is? Current trends towards an open definition. Computer Communications. 2017;**109**:117-130

[24] Guevara JC, Bittencourt LF, da Fonseca NLS. Class of service in fog computing. In: Proceedings of the 2017 IEEE 9th Latin-American Conference on Communications (LATINCOM), Guatemala City, Guatemala. IEEE; 2017. pp. 1-6. DOI: 10.1109/LATINCOM.2017.8240187

[25] Vilalta R, Lopez L, Giorgetti A, Peng S, Orsini V, Velasco L, et al. TelcoFog: A unified flexible fog and cloud computing architecture for 5G networks. IEEE Communications Magazine. 2017;55:36-43

[26] Kim W, Chung S. User-participatory fog computing architecture and its management schemes for improving feasibility. IEEE Access. 2018;**6**:20262-20278

[27] Kaur M, Bharti M. Securing user data on cloud using fog computing and decoy technique. International Journal of Advance Research in Computer Science and Management Studies. 2014;**2**(10):104-110

[28] Bonomi F. Connected vehicles, the internet of things, and fog computing. In: The Eighth ACM International Workshop on Vehicular InterNetworking (VANET). Las Vegas, USA: Sigmobile; 2011

[29] ETSI. ETSI and Open Fog Consortium Collaborate on Fog and Edge Applications. 2017. Available from: https://www.etsi.org/newsroom/news/1216-2017-09-news-etsi-and-openfog-consortium-collaborate-on-fog-and-edge-applications

[30] Petri I, Rana OF, Bignell J, Nepal S, Auluck N. Incentivizing resource sharing in edge computing applications. In: Proceedings of the International Conference on the Economics of Grids, Clouds, Systems, and Services (GECON'17). Springer Link; Oct 2017. pp. 204-215. Available from: https://link.springer.com/chapter/10.1007/978-3-319-68066-8_16

[31] Bravo C, Backstrom H. "Edge computing and deployment strategies for communication service providers," White Paper GFMC-20:000097, 2020. Available from: https://www.ericsson.com/en/reports-and-papers/white-papers/edge-computing-and-deployment-strategies-for-communication-service-providers

[32] Badreddine W, Zhang K, Talhi C. Monetization using Blockchains for IoT Data Marketplace. Montreal, Quebec, Canada: Ecole de Technologie Superieure; [Online]. Available: wiem.badreddine.1@ens.etsmtl.ca, kaiwen.zhang@etsmtl.ca, chamseddine.talhi@etsmtl.ca

[33] Poggi A, Tomaiuolo M. Mobile agents: Concepts and technologies. In: Handbook of Research on Mobility and Computing: Evolving Technologies and Ubiquitous Impacts. Beijing, China: IGI Global; 2011. pp. 343-355. Available from: https://www.igi-global.com/chapter/mobile-agents-concepts-technologies/50597

[34] Nguyen D-D, Ali MI. Enabling on-Demand Decentralized IoT Collectability Marketplace Using Blockchain and Crowdsensing. In: Proceedings of the IEEE. IEEE; 2019

[35] Debe M, Salah K, Rehman MHUR, Svetinovic D. Monetization of services provided by public fog nodes using Blockchain and smart contracts. IEEE Access. 2020;**8**:20118-20128

[36] Yu Y, Liu S, Guo L, Yeoh PL, Vucetic B, Li Y. CrowdR-FBC: A distributed fog-Blockchains for Mobile crowdsourcing reputation management. IEEE Internet of Things Journal. 2020;**7**(9):8722-8735

[37] Rejiba Z, Masip-Bruin X, Marín-Tordera E. Analyzing the deployment challenges of Beacon stuffing as a discovery enabler in fog-to-cloud systems. In: Proceedings of the 2018 European Conference on Networks and Communications (EuCNC), Ljubljana, Slovenia; 2018. pp. 1-276. DOI: 10.1109/EuCNC.2018.8442471. Available from: https://ieeexplore.ieee.org/document/8442471

[38] Zhang Y, Zhang H, Long K, Zheng Q, Xie X. Software-defined and fog-computing-based next generation vehicular networks. IEEE Communications Magazine. 2018;**9**:34-41

[39] Cao B, Zhang L, Li Y, Feng D, Cao W. Intelligent offloading in multi-access edge computing: A state-of-the-art review and framework. IEEE Communications Magazine. 2019;**57**(3):56-62

[40] Cui Q, Gong Z, Ni W, Hou Y, Chen X, Tao X, et al. Stochastic online learning for mobile edge computing: Learning from changes. IEEE Communications Magazine. 2019;**57**(3):63-69

[41] Kim D, Lee H, Song H, Choi N, Yi Y. Economics of fog computing: Interplay among infrastructure and service providers, users, and edge resource owners. IEEE Transactions Mobile Computations, to be Published. 2020;**19**:2609-2622. DOI: 10.1109/tmc.2019.2925797

[42] Li X, Zhang C, Gu B, Yamori K, Tanaka Y. Optimal pricing and service selection in the mobile

cloud architectures. IEEE Access. 2019;7:43564-43572. DOI: 10.1109/access.2019.2908223

[43] Vega F, Soriano J, Jimenez M, Lizcano D. A peer-to-peer architecture for distributed data monetization in fog computing scenarios. Wireless Communications and Mobile Computing Volume. 2018;**2018**, Article ID 5758741:15 https://doi.org/10.1155/2018/5758741

[44] Caprolu M, Di Pietro R, Lombardi F, Raponi S. Edge computing perspectives: Architectures, technologies, and open security issues, 2019 IEEE international conference on edge computing. IEEE Access. 2020;**8**:231825-231847

[45] AbdulKareem KH, Mohammed MA, Gunasekaran SS, Al-Mhiqani MN, Mutlag AA, Mostafa SA, et al. A review of fog computing and machine learning: Concepts, applications, challenges, and open issues. IEEE Access. 2019;7:141274-141308

Chapter 6

Design and Implementation Smart Petrol Station Using Internet of Things

Zahra'a M. Baqir and Hassan J. Motlak

Abstract

The problem that still exists nowadays with the petrol station is the method of operation because the petrol station is currently operated manually. This chapter adopted an automatic fuel filling algorithm based on a Radio Frequency Identification (RFID) card that carries information about the user regarding the type of user, the available balance and the permitted amount of fuel. This data is sent via the Wi-Fi module (ESP-32S) to the IoT platform (the panel), where it is encrypted with the hashing function and the Advanced Encryption Standard (AES) algorithm before being displayed by authorized access to the platform dashboard, which is It is also encrypted with an ID and a secret number.

Keywords: Internet of Things (IoT), RFID, ESP32 microcontroller, sensors, ThingsBoard platform, petrol station

1. Introduction

Regarding to this research work, the research work will be divided into three categories, In first part, petrol station hardware which consist of an NFC RFID reader for cashless payment, sensors are going to be used in measurement of the petrol transferred to the customer's tank, levels and the temperature of fuel in storage tank and sensors for fire protection and environmental monitoring. Calculations and programming are to be made through at Arduino mega 2560 and ESP-32S microcontrollers. For the second part, the ESP-32S module is used to enable wireless communication interface between the petrol station hardware system and the third part, which is the petrol station monitoring system, which is represented by the IoT ThingsBoard server that is downloaded to Raspberry Pi as a local server for data collection, processing, visualization, and system management. Arduino programming language is used in order to create a systematic software as the main control system for this petrol station system.

2. Literature survey

2.1 RFID-based smart petrol station

Implementation of a smart petrol station system in several countries can be read in many literatures all these projects used RFID tag as a fee card.

In 2016, Punit Gupta, et al. [1] described the design and implementation of a smart fuel station that will calculate the amount of fuel in the petrol station and send the information to a main server. They apply ultrasonic sensor to measure the level of petrol.

In 2016, Gowri Shankar, et al. [2], the proposed Biotelemetry system is used in this research to confirm the singular user with their Petrol tag. Biotelemetry is being employed as a thumbprint sensor in this case.

In 2016, Wavekar Asrar, et al. [3], had the project's major goal is to create a system that uses RFID technology to automatically deduct the amount of gasoline dispensed from a user's card. The use of RFID technology to control fuel dispensing in Indian cities is proposed in this study.

In 2017, Kumaresan and Babu Sundaresan [4] presented paper, it allocates with automation of a petrol station selling outlet; this system will provide the transactions and supply report to the holder for each hour. In this thesis, the user can access the present status of the petrol station as well as the stock maintenance through the net request.

In 2018, Chandana et al. [5], had the goal of this project is to build a security mechanism for filling gasoline at petrol stations without involving humans. The RFID smart card eliminates the risk of carrying cash at all times and also has a prepaid recharging option.

In 2018, Dongarsane et al. [6], had the secure and atomized delivery of gasoline is making possible by the simple and proper use of microcontroller and GSM technologies. It comes with an easy-to-use mobile phone system and a colorful user interface (GUI).

In 2019, Rashmitha [7] suggested a project where the customer has a smart card in this project. The card is nothing unless it has a magnetic member inserted in it. When a consumer presents this tag to the reader, the reader reads the unique number and sends the sign to the microcontroller.

In 2019, Nitha Velayudhan et al. [8], presented study proposes an automated petrol station management system that addresses the shortcomings of the current system. They used GSM and AT Mega 328 to develop an automatic gasoline filling system. This solution has the potential to make the fueling process more convenient, reliable, and secure.

Compared with the projects presented in the above sources, it is clear that our designed project has combined a number of mechanisms used in the construction of a smart fuel filling station. In our project, the researchers used the Arav De card to deal with secure electronic payment. The petroleum card has also been distinguished with a unique number that distinguishes between the type of users and thus controlling the distribution of fuel in Iraqi cities.

2.2 Networked petrol station systems for the IoT

In 2017, Sahana Rao et al. [9], used RFID and GSM technology to automate the filling of fuel at a petrol station. The transactions are designed to be user-friendly, which means that they are available at the touch of a button on the customer's smartphone.

In 2017, Carlo Makdisie and Badia Haidar [10], produced the stated system by assistance of present programs (Labview, Protues, PIC C Compiler) to reach a high precise control of the needed parameters, furthermore to using SCADA for managing and controlling all the system to escape any sudden faults; similar fire disaster and building the system work with extreme precision.

In 2018, Deepa et al. [11], used GSM and RFID, this paper created an automated fuel pump. All users in this system have an RFID card, which can be charged via a few different points. The fuel station is prepared with a smart card reader that detects the amount in the card as well as all security information and displays it on the LCD.

In 2019, Pranto and Rahman [12] presented paper indicates a design in which a system will be capable to trace a vehicle's petrol refill amount, cost & time then notifies the vehicle's owner about the refill as well. The system is divided into two parts: A mobile application and a petrol station.

In 2021, Sembodo and Atmajaya [13] presented monitoring tool uses the Arduino ESP8266 which is combined with Internet of Things (IoT) Technology which is displayed on the LCD and the Internet Web. Customers can purchase the patrol at gas stations independently, by connecting the smartphone application with the gas station wireless and purchase and make payments through the application.

In our designed project, three filling stations were connected, each one controlled by the Arduino Miga and ESP32S, and connected to them by sensors to represent the lower layer (sensing layer) of the IoT infrastructure. Data is transferred to the network layer using the Message Queuing Telemetry Transport)MQTT(protocol to represent the middle layer. The data from this layer is sent to the local server (Raspberry Pi) or a cloud server. The data is displayed in a web application from the IoT platform, ThingsBoard.

3. The proposed system

In this chapter, we proposed a design system of three smart petrol stations that are monitored and managed remotely from a single interface on IoT platform. a smart petrol station system that performs each of the following tasks: Detecting the fuel level inside the storage tank, measuring the fuel temperature, an automatic fuel dispenser, electronic payment, monitoring the environment through the temperature and

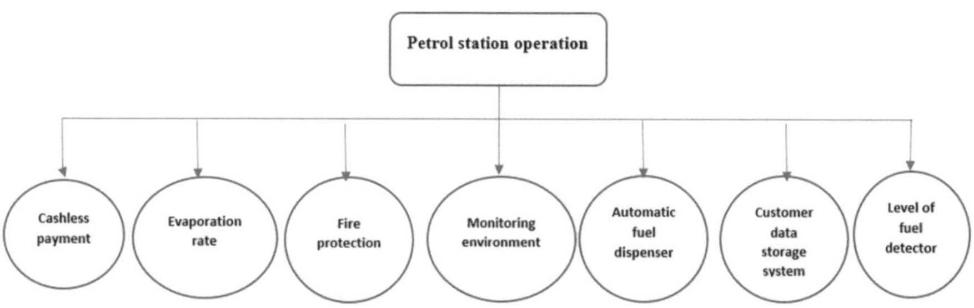

Figure 1.
Functions of the proposed smart petrol station as IoT application.

humidity sensor, detecting fires and extinguishing them automatically and storing all related data of customers and station with a database stored on a local server or applied to a cloud server. This data can be accessed, viewed and processed through the ThongsBoard web application, which can be accessed from any web browser on smart devices or desktop. **Figure 1**, shows the operations of the proposed system.

3.1 Architecture of the proposed system

Figure 2, shows the block diagram of designed system of three smart petrol stations based on IoT. The petrol station system connects to local server or cloud server over the internet. Online data can be visualized by a remote company office on ThingsBoard web user interface through chrome or Firefox web browser from desktop or smart phone or other devices. **Figure 3**, shows the flow chart of this system. The scheme offered consists of hardware part and software part.

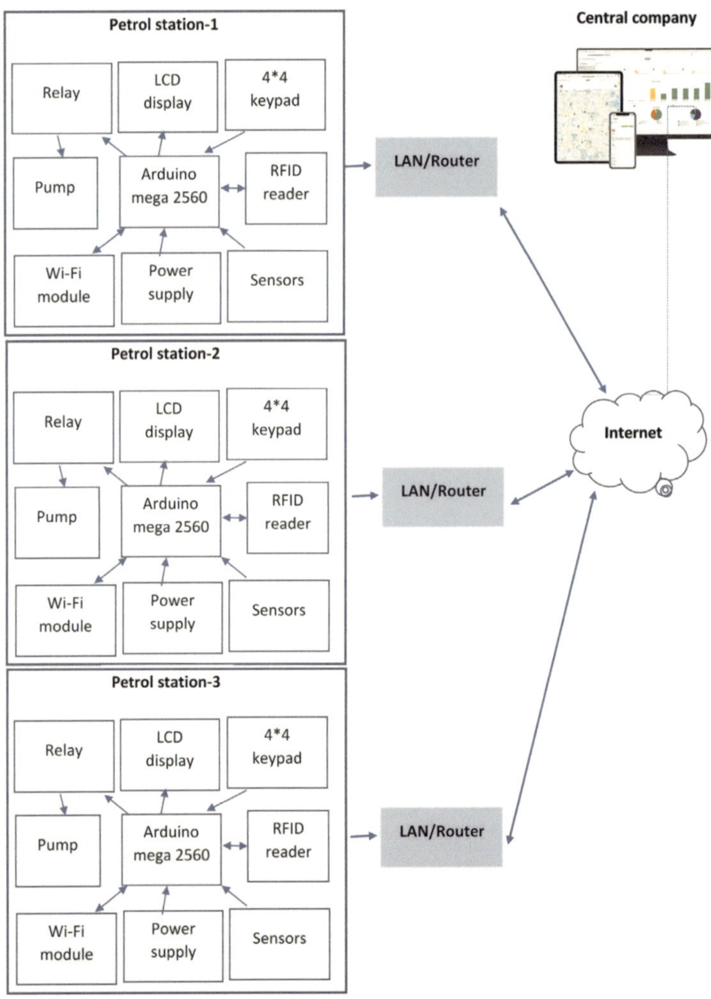

Figure 2.
Block diagram of the proposed system.

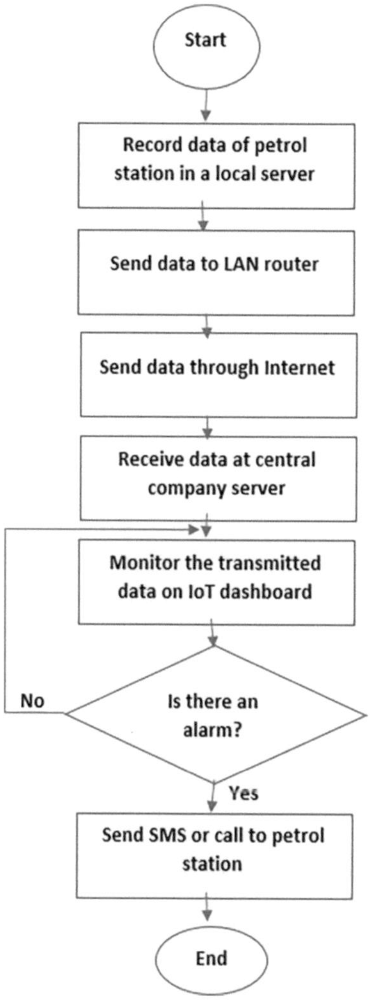

Figure 3.
Flow chart of the proposed system.

3.1.1 Hardware part

It consists of four layers as IoT architecture. These layers are:

3.1.1.1 Sensing layer

This layer is represented by the control system that controls the fuel dispenser and storage tank.

3.1.1.2 Gateway layer

This layer is represented by the Wi-Fi router, which receives the sensor data from the station and sent by ESP-32S via the MQTT protocol to be directed to the server at the service layer.

3.1.1.3 Service layer

The service layer of the system designed from the server consists of a Raspberry Pi 4 model B minicomputer with a 1.5 GHz quad-core 64-bit processor that provides about 3 times better performance compared to the previous version, more ports and faster Internet connection, where the certified Raspbian operating system is downloaded [14]. On the Linux system, and downloading the ThingsBoard platform on this system for the purpose of reviewing all the variables and operations that the customer performs, and they occur at the packing station immediately after logging into the server by entering a fixed IP address in any web browser such as Chrome, Firefox or others. Or dealing with cloud ThingsBoard Server for the purpose of accessing station information from anywhere in the world via the Internet. For more security this server can be accessed using a username and password to view data in the application layer.

3.1.1.4 Application layer

It uses a user interface where the central company managing the three stations can view data from any station using any web browser from a desktop or smart phone.

3.1.2 Software Part

The petrol station is controlled by a system to enable that it operates efficiently. As this petrol station is expected to be left standalone and managed wirelessly, a system to control and monitor it from a distance is necessary. Software that was used in this application are:

3.1.2.1 Arduino Software (IDE)

The Arduino Integrated Development Environment (IDE) is a free and open-source cross-platform application (for Windows, macOS, and Linux) developed in C and C++ functions. The Arduino IDE makes it simple to develop code and upload it to the device [15]. This software can be used with any Arduino board. Arduino programming is used to control the petrol station system. The Arduino Mega is communicated with the ultrasonic sensor for the fuel leveling and with other sensors in the project for environmental monitoring and controlling. Aside from that, Arduino programming controls the 4 x 4 keypad number, RFID system, and LCD display that serves as the customer's interface. Flow meter and relay are responsible for determining the functionality of the pump, needed to be programmed by Arduino. Finally, Arduino IDE uses the ESP-32S module in programming to establish a connection with the internet, in order to provide these functionalities;

- Determines sales and store a database for the petrol station on local server.
- Monitors the temperature and humidity of weather.
- Provides information about the petrol level (stock) to the server.
- Provides information to the server about evaporation and density of fuel in the storage tank (stock).

3.1.2.2 Proteus program for simulation

"Virtual System Modelling" (VSM) by Proteus combines SPICE simulation in mixed modes with the industry's fastest microcontroller simulation. It allows for quick hardware and firmware prototyping in software [16]. Proteus VSM's most significant aspect is its ability to simulate the interface among software functioning on a microcontroller and any digital or analog devices linked to it. The microcontroller simulation is put on the schematic among the other parts of design of your product. It mimics object code execution in the same way as an actual chip does. Logic levels in the circuit will change whenever your program code writes to a port, and if the circuit varies the situation of the pins of processor, your program code will notice, just like in physical life. The VSM CPU models accurately emulate each supported processor's input/output ports, timers, interrupts, USARTs, and other components. Because the interface of each of these peripherals with the outside circuit is thoroughly described to the level of waveforms, the whole system is being tested, it is far more than a simple software simulator. With over 750 microprocessor variations supported, a lot of fixed SPICE models, as well as one of the major libraries of entrenched simulation peripherals in the world, for an embedded model, Proteus VSM is still the best option.

3.1.2.3 LabVIEW program for simulation

LabVIEW (Laboratory Virtual Instrumentation Engineering Workbenches (where a national instruments platform and development environment for a visual programming language. The goal of this type of programming is to automate the use of laboratory decision-making and measurement equipment. It's a proprietary software package that makes it easier to design different types of data management, control, and measurement systems and is therefore ideal for GUI design.

3.2 Implementation of the proposed system

The designed system is divided into two main parts: the first part relates to the hardware on the board (control system) and the second part relates to the ThingsBoard web server. In **Figure 4**, the arrows pointing toward the Arduino mega 2560 shows the input, while the arrows pointing from Arduino shows the output of the processor. For a single petrol station two parts are implemented as shown in **Figure 4**.

First: the ultrasonic sensor HC—SR04 connected at the top of a fuel tank to calculate the fuel level by sending ultrasound waves. The distance is equal to the speed multiplied by the time, the speed of sound is known at a temperature of 20 degrees Celsius equivalent to 343 meters/second. The time that the sensor processor calculated from the start time of the pulse to the echo return time is divided by 2. By knowing the time and speed, it is possible to calculate the distance, determine the height level, send the data to the microcontroller (ATmega2560) connected to the ESP32 Wi-Fi module and send the data later to the server. A fluid temperature sensor (DS18B20) has been added to the tank, and it can be added to the pump hose to measure the temperature of fuel and the consequent result of calculating is the evaporation rate and density which dependent on temperature to determine types of mixtures. DS18B20 connects via 1-WIRE protocol requiring one data line to connect to the Arduino Mega Board.

Second: the fuel dispenser system is a box where the electronic parts appearing on it, which represented by a 4 * 4 keyboard, have been linked. The letters on the panel

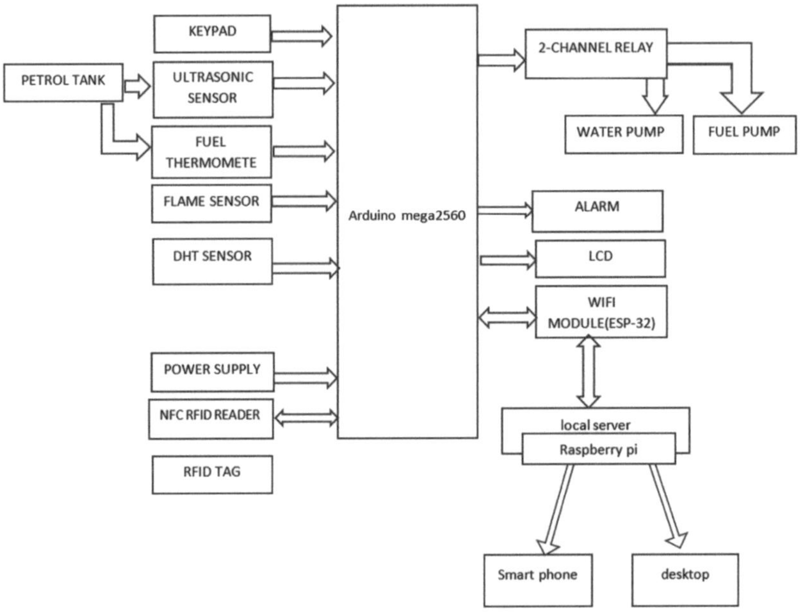

Figure 4.
The proposed smart petrol station.

perform a number of functions. Also, a 20 * 4 LCD screen was attached to the box to show the customer all the functions when pressing the keypad. NFC RFID to scan the RFID tag by establish a wireless connection with each other by touching or bringing them close at close range, usually not more than a few centimeters. NFC RFID and LCD communicate with each other and send data between them via the (I2C) protocol, which is a serial data communication protocol, which operate among two devices or a group of devices to transfer information among them. The protocol works in a "half-duplex" manner, one of the devices will be sending whereas the another will be at the receiving state and they cannot transmit at the same time. The DHT11 temperature and humidity sensor has been added to the contents box to monitor the temperature and humidity to be an example of monitoring the environment and showing the results directly on the server from the second petrol station. The fuel and water pumps are connected on the sides of the box; the fuel pump is connected to the fuel tank on the one hand and linked to a flow meter that calculates the number of liters required; the flow meter is connected to a hose on top of which hose there is a button that can be pressed to fill the required amount of fuel; and if the required quantity is filled, the pump will revert to its original position. As for the water pump, it has been linked to a flame sensor that sounds an alarm when a flame is detected, as it sends its data to the microcontroller that instructs the relay switch associated with the water pump to operate it. This pump can also be controlled manually via a push button and the pump is started. **Figure 5**, shows the connection map establishes by fritzing software for the hardware parts on the board. **Figure 6**, shows a section of the practical hardware of the designed and implemented system, which is installed outside the control system box.

Inside the control system box, there is an Arduino Mega 2560, a microcontroller board based on the ATmega2560 processor. The board contains everything needed to support the microcontroller; It can be connected to a computer with a USB cable or powered by an AC-to-DC adapter or battery to start. The ATmega2560 contains 256 KB

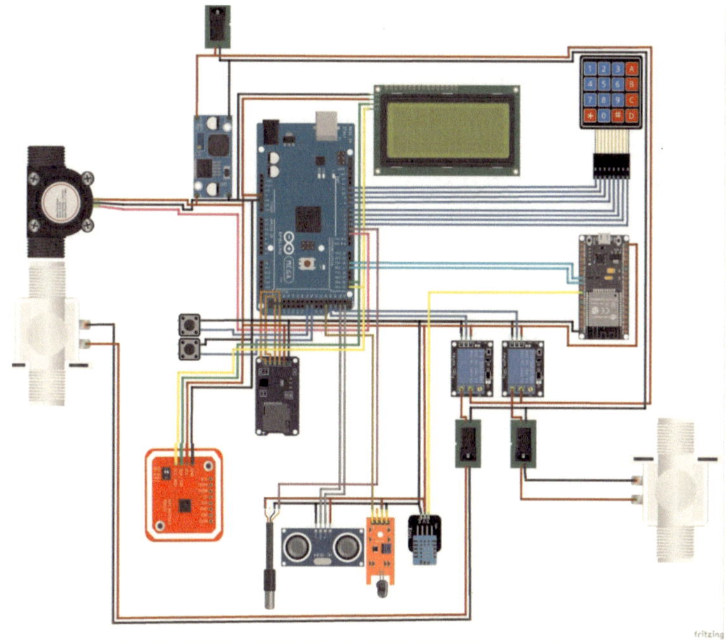

Figure 5.
Connection map on the board.

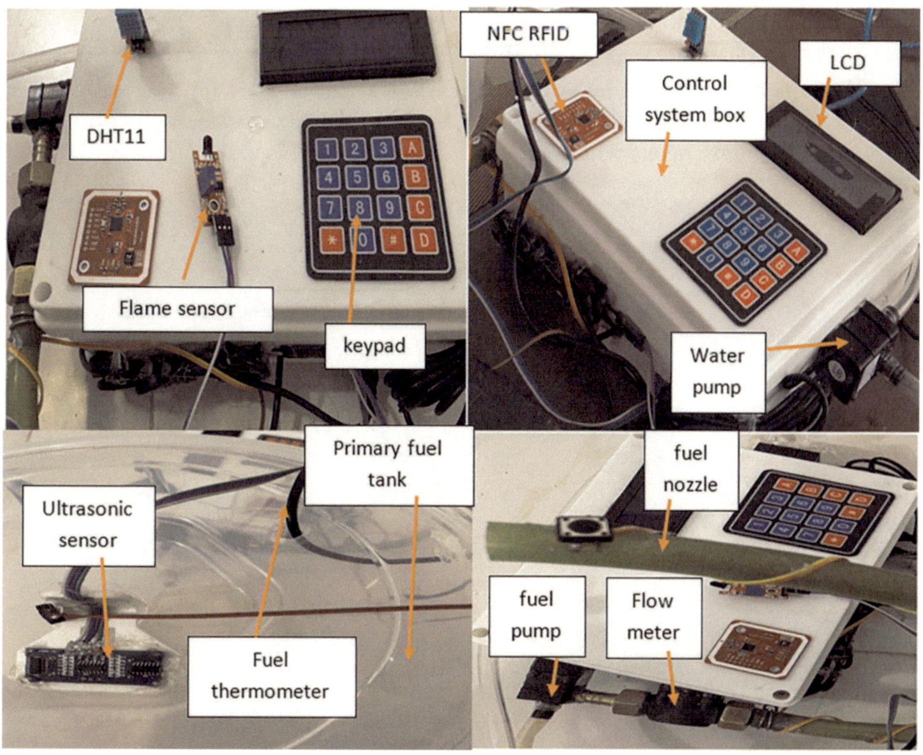

Figure 6.
Hardware section outside the box of the control system.

of flash memory to store the code (of which 8 KB is used for the bootloader), 8 KB of SRAM and 4 KB of EEPROM. And because of this memory is little, the MicroSD card adapter module, which is a Micro SD card reader for reading and writing through the file system, and the SPI driver, has been linked, so this module connects with the microcontroller through the SPI serial communication protocol. This protocol works through three lines "MISO, MOSI & SCK". At any given time, only two devices can talk to each other. What distinguishes this protocol is that it is (Full-duplex) that is, the two devices can transmit and receive at the same time. Also, from the electronics that were provided for this system inside the box, the voltage regulator is fixed, as it has a voltage of 12 volts to feed the pumps and a voltage of 5 V comes out of it to feed the rest of the system elements. The two channel relay module MR009–004.1 is fixed inside the box, where is a module designed with two relays that provide 12 volt power to allow control of the fuel pump and the water pump. WiFi-BT-BLE MCU module the core of this module is an ESP32 chip that supports Secure Over-The-Air (OTA) upgrade. This module inside the box is responsible for providing a direct connection to the server via a WiFi router and it acts as a second microcontroller represent as second petrol station. **Figure** 7, shows the core components of the control system box.

3.2.1 Methodology of the proposed system in proteus simulator

In Proteus software, the methodology of the system designed for a single filling station has been simulated locally to simulate the following functions:

3.2.1.1 Fuel Filling

The NFC RFID detects the ID number of RFID card and sends it to the Arduino Mega 2560 microprocessor when the RFID card or tag comes within range of the RFID reader. The ID codes of the RFID cards are already recorded in the microcontroller. The microcontroller checks whether or not this RFID card is valid. If the RFID card is

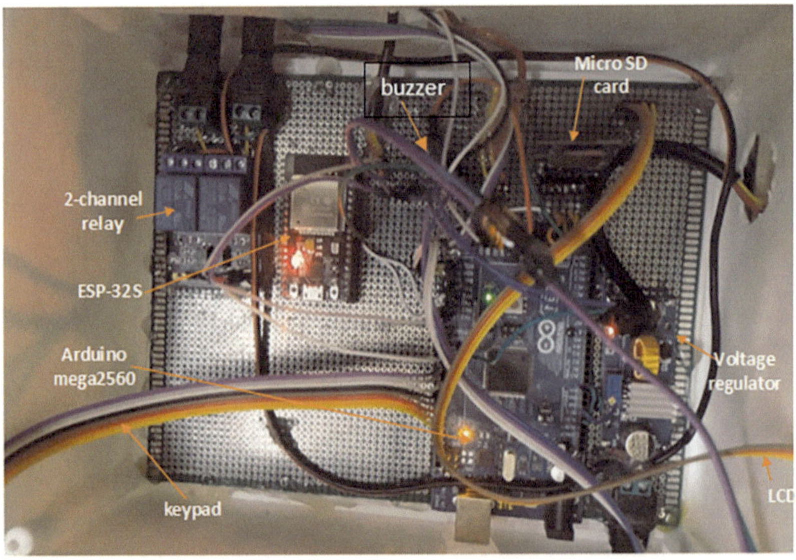

Figure 7.
Inside the control system box.

authorized, the microcontroller displays the RFID card user on the LCD screen after a check; otherwise, nothing appears, indicating that the RFID card is unauthorized. If the RFID card is authorized, the microcontroller allows the consumer to enter the amount of gasoline via the keypad, which is then displayed on the screen. The microcontroller then checks to see if the RFID card's available balance exceeds the amount entered. If there are more than that, the microcontroller activates the fuel pump motor through a relay switch for a set amount of time before turning it off. If the user wants to examine the quantity of petrol dispensed and the remaining balance of the RFID card after the procedure is completed, the information on the amount of petrol dispensed and the remaining balance of the RFID card is displayed on the LCD screen. Algorithm of automatic fuel filling pump shown in **Figure 8**. All this information about users is visualized to the IoT ThingsBoard server which installed on raspberry pi as local server, using Wi-Fi module ESP-32S. On a local server, we keep track of the amount of gasoline distributed by users, which is protected by a password that is only known by the fuel company. If the customer's available balance is less than the amount entered, the RFID card can be recharged. In this case, the RFID card functions as a prepaid card that may be recharged at the fuel station.

3.2.2 Methodology of the proposed system in LabVIEW

In LabVIEW software, the methodology of the system designed for a three-filling station has been simulated locally to simulate the following functions: level of fuel in

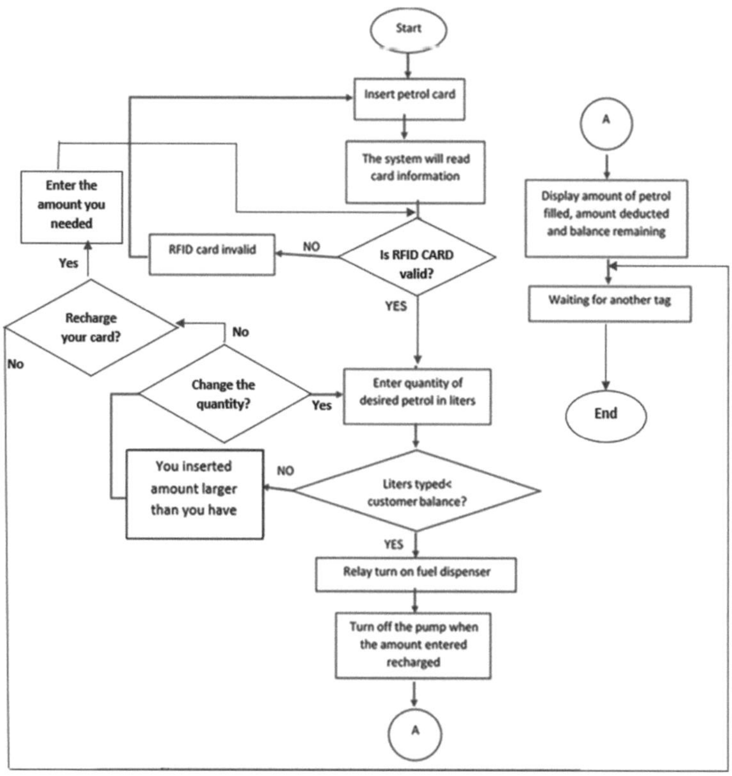

Figure 8.
Algorithm of the automatic fuel filling pump.

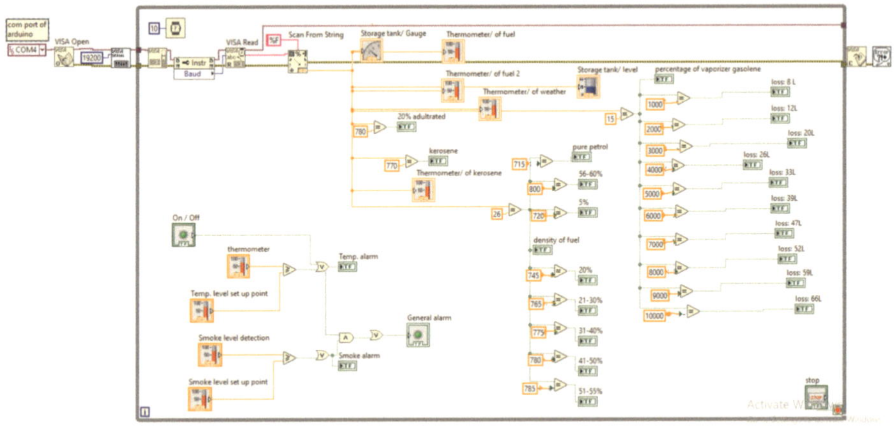

Figure 9.
Smart petrol station via LabVIEW.

the storage tank, the temperature of the fuel temperature and humidity of weather and fire alarm. As shown in **Figure 9**.

4. Simulation results

Simulation results of the system designed and implemented by Proteus and LabVIEW programs were achieved.

4.1 The results in proteus

The design suite Proteus 8.9 is an exclusive software implement suite that was principally used in this project for electronic project automation. In Proteus, a hex file to the microcontroller portion on the diagram was used to operate the Arduino mega2560 simulation. It is next co-simulated along through whichever digital and analog electronics linked to it. In **Figure 10**, we build a prototype by selecting a few blocks and using the Simulink block toolbar to join them together. Results were achieved on how to perform each of the following:

4.1.1 Monitoring and controlling storage tank

The spread of USTs represents an actual risk that requires difficult and costly methods for its management. But by adopting the method used in the simulation, it is possible to control and monitor the level of petrol available in the storage tanks in an easy and inexpensive way. **Figure 11**, shows the design of the power supply is simple and cheap, when it is turn ON. First, Arduino mega2560 checks the petrol level by ultrasonic sensor at the top of the tank. if a petrol level is low than threshold value, a buzzer is initiated and displayed low petrol level on the LCD and Arduino mega operates relay which connected to run DC motor of a fuel supply pump to supply fuel to the tank. In **Figure 12**, shows when a fuel level rises above the

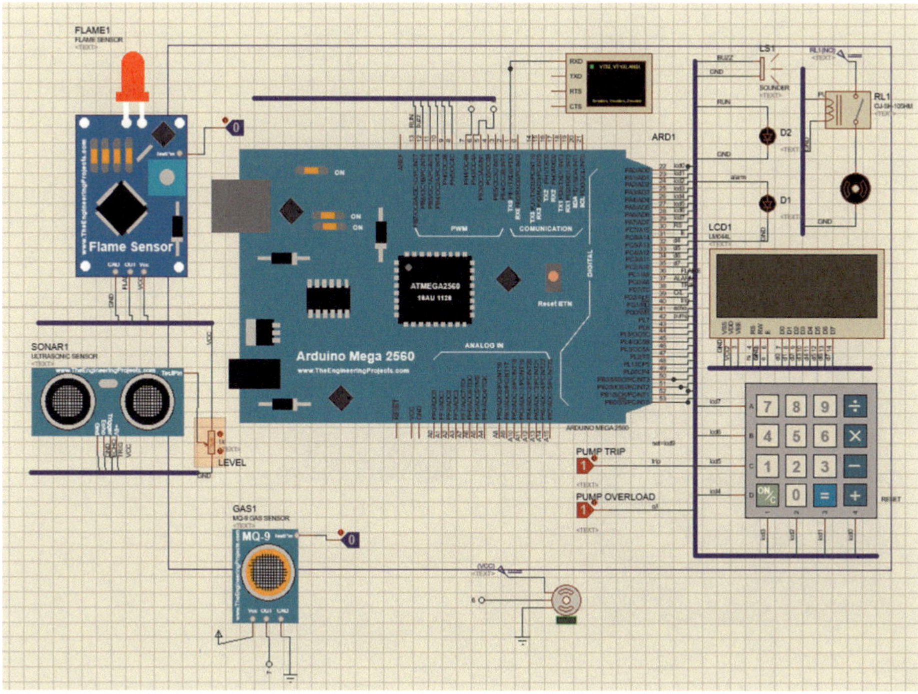

Figure 10.
Initialized subsystem before launched to IoT server in Proteus.

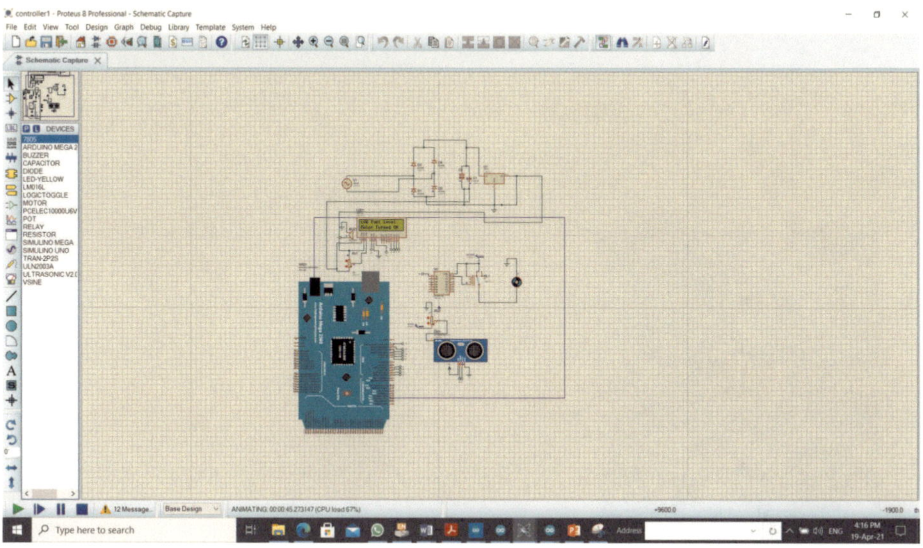

Figure 11.
Simulation of the management storage tank (low fuel level).

threshold value, the buzzer is turned off and levels of fuel displayed on LCD screen. When the fuel level reaches the full level as set in the programming, Arduino mega turns off the relay which connected to stop the dc motor of fuel supply pump running,

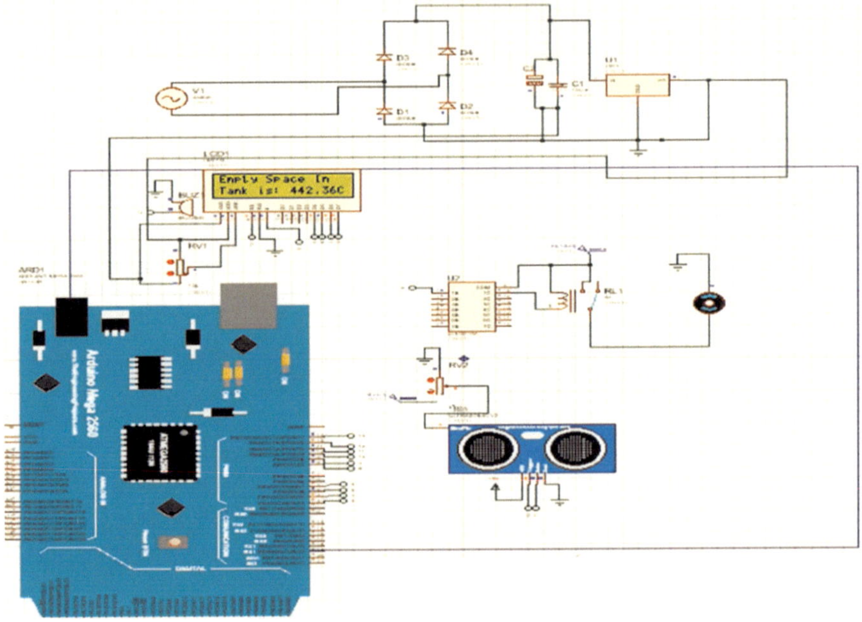

Figure 12.
Simulation of the management storage tank for the proposed system (display fuel level).

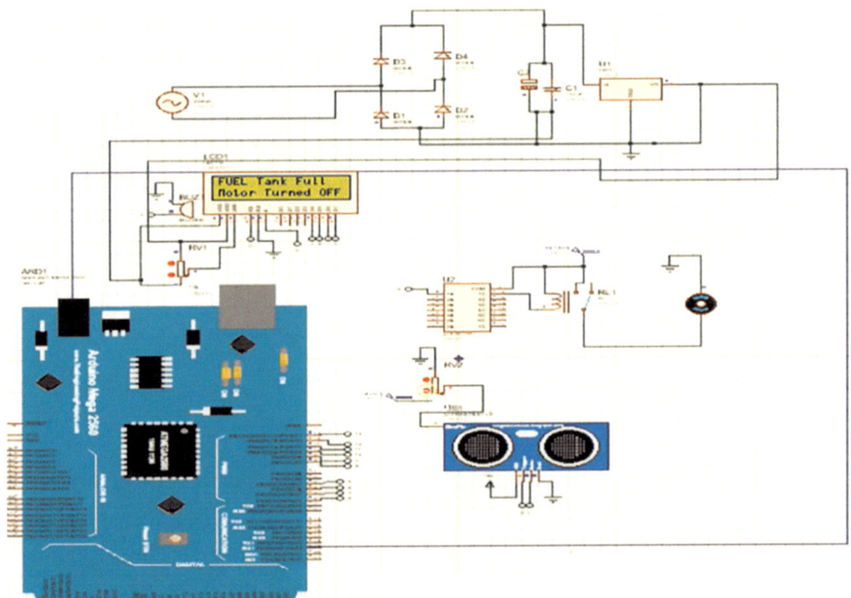

Figure 13.
Simulation of the management storage tank for the proposed system (full fuel tank).

the LCD screen displays that the tank is completely full as shown in **Figure 13**. **Figure 14**, shows how the fuel levels in the storage tanks can be displayed on the serial monitor.

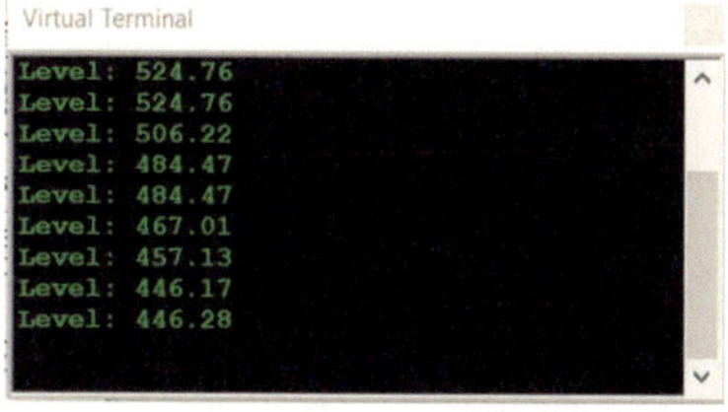

Figure 14.
Fuel level of the storage tank on a serial monitor.

4.1.2 A fuel filling operation

Sub-system simulation of the proposed system in the Proteus program, to reveal how the system performs in the fuel filling process. The result of the subsystem, along with a step-by-step approach, is listed below:

Step 1: Swipe the RFID card and type in the ID number. Where it appears from the **Figure 15**, that Arduino mega2560 connected to the other parts of the subsystem, the green LED is activated to indicate that the petrol station is operating normally, which are of interest to us in this step, the connection between the microcontroller and Ultrasonic sensor installed at the top of the fuel tank, which measures the height of the fuel and send its data to Arduino mega to display the fuel level on the LCD screen, as for the benefit of connecting the microcontroller to the keypad, it appears by entering the ID number of the card swiped on the virtual card reader as shown in **Figure 16**.

Step 2: If RFID card is authorized, then the controller fetches the user available amount data from server and display on LCD, then microcontroller ask the user to enter the amount of petrol as shown in **Figure 17**.

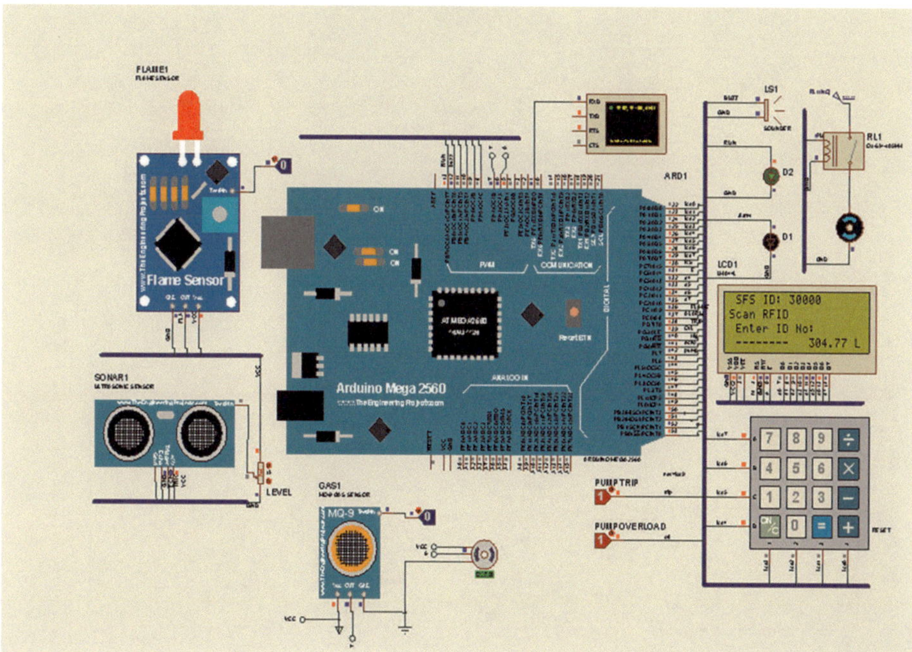

Figure 15.
LCD shows level of fuel tank.

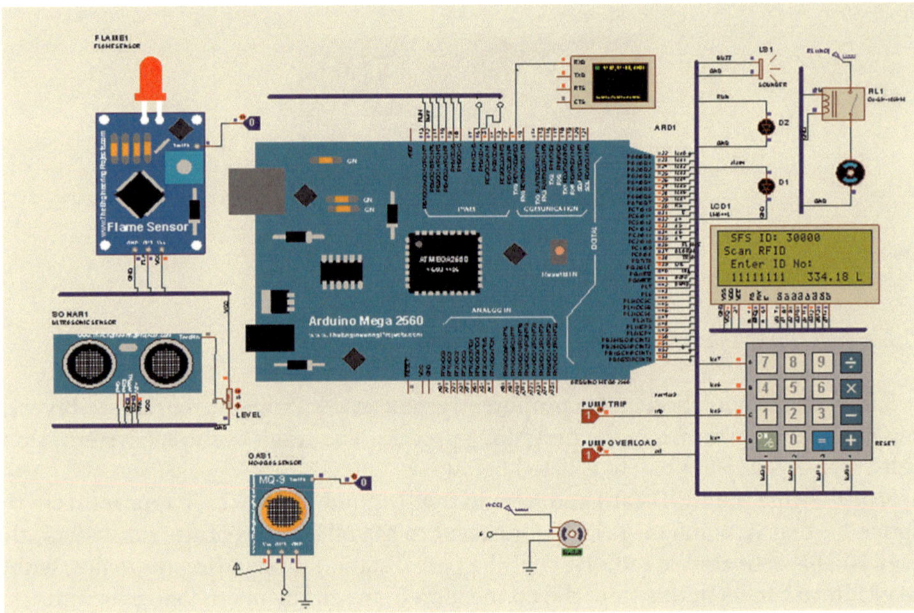

Figure 16.
Enter ID No by keypad.

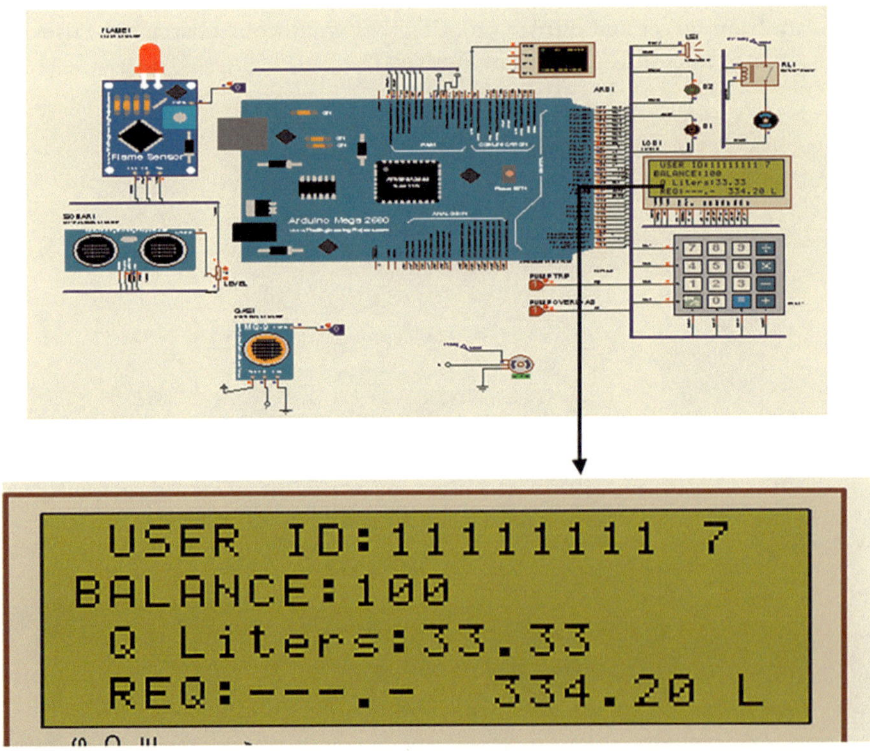

Figure 17.
User ID and balance.

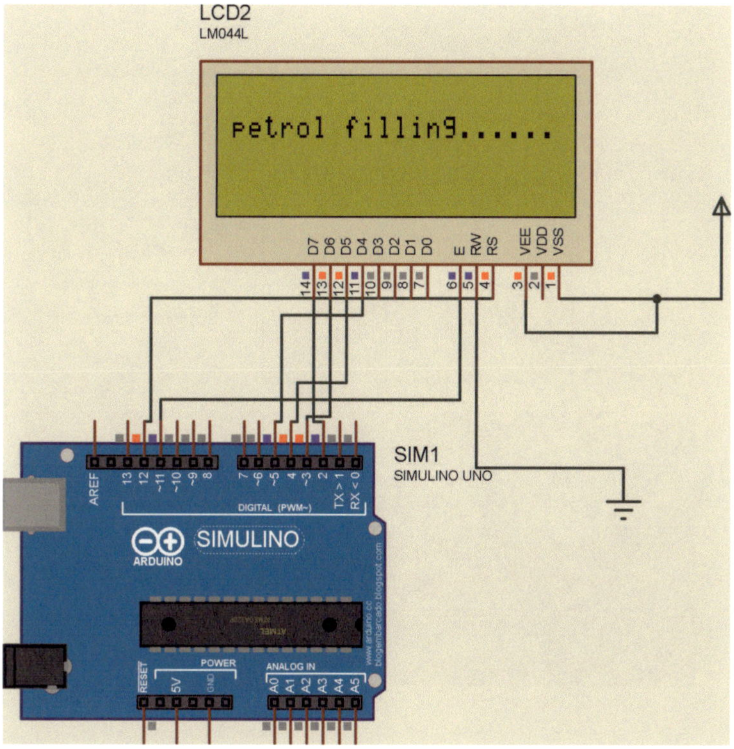

Figure 18.
The petrol filling process has started.

Step 3: If an entered amount is less than the available amount, then petrol filling process starts (i.e., Motor turns ON) as shown in **Figure 18**. Otherwise microcontroller say to recharge a suitable amount as shown in **Figure 19**.

Step 4: After filling the petrol, the petrol filled amount and available balance information is display on LCD screen as shown in **Figure 20**. And available balance is updated in the server.

4.1.3 Monitor fire protection

The microcontroller in this system is connected to a flame sensor and servo motor as a fire extinguisher (that discharges a jet of water, foam, gas, or other material to extinguish a fire). When there is a fire, the flame sensors detect it, then the warning system - indicated by the buzzer and red LED - is activated, and the servo motor is activated to convert its angle of 180 degrees. When the alarm state ends, the servo motor returns to its original position, and the LED and buzzer turn off. As seen in **Figure 21**, a fire alert shows on the LCD screen.

4.1.4 Monitor security of petrol station

Safety of fuel pump also tested in the occurrence of a buildup in the load, to indicate a warning situation on the LCD screen, the buzzer and the red LED were

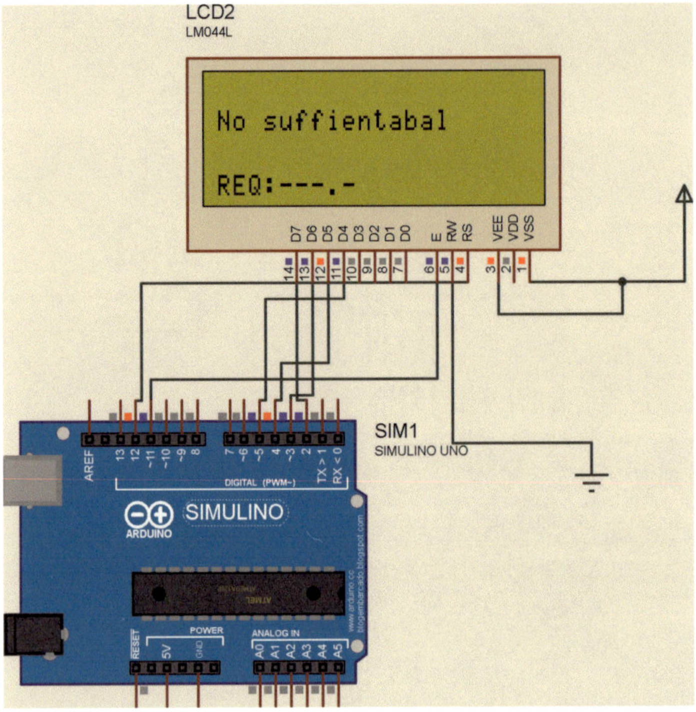

Figure 19.
If the entered amount is more than the available amount.

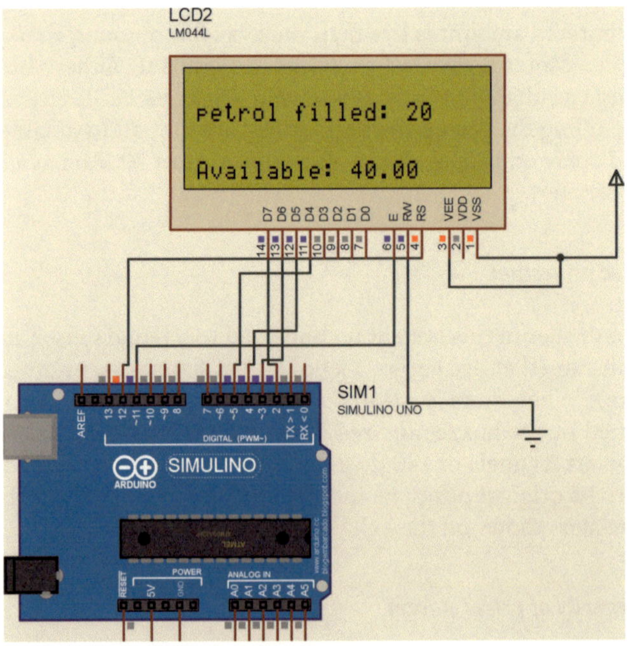

Figure 20.
The amount of petrol loaded and the available balance.

Figure 21.
Simulation of management fire protection of proposed system.

triggered to tell the viewer of the status of the load besides work to switch off the relay that runs the pump motor, as displayed in **Figure 22**.

4.2 The results in LabVIEW

The designed strategy is implemented in the LabVIEW VI environment, and simulation results are presented to validate the proposed controller in **Figure 23**. Where it appears from Figure monitoring the central station of three stations, the design assumes the following in each station:

1. Monitoring the fuel level inside the storage tanks for three products. Where a visual alert appears represented by a green LED indicating an acceptable level of fuel rise inside the storage tank, while if the fuel level drops below 20%, a visual alert represented by a red LED appears indicating the necessity of refilling the fuel to the indicated tank.

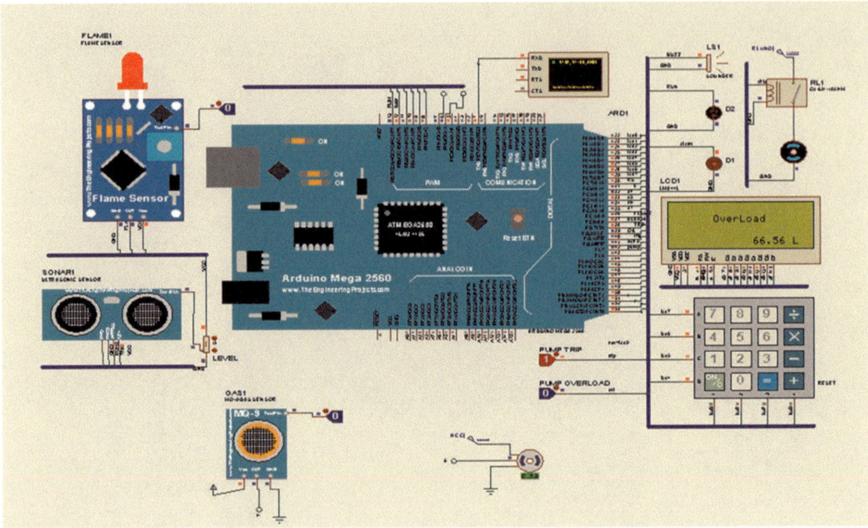

Figure 22.
Overload status.

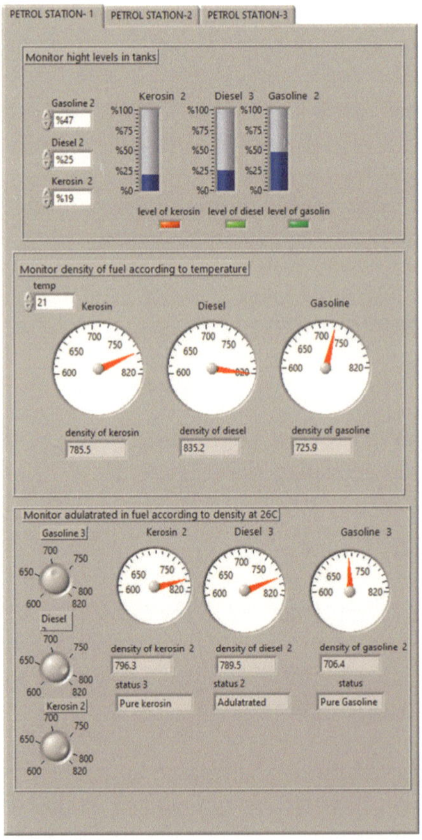

Figure 23.
The output parameters from LabVIEW.

2. Monitoring the density of the fuel according to the temperature. With a temperature of 21 Co, the density of gasoline was recorded at 725.9, while the density of diesel was 835.2, and the density of kerosene was recorded at 785.5

3. Quality control of the product according to the density: with a temperature of 26 Co, the density of gasoline was 706.9, and at this density the textual reference to the purity of the petroleum is made. At a density of 785.5 for diesel, the presence of impurities mixed with diesel was indicated, while kerosene at a density of 796.3 indicated the purity of kerosene.

In **Figure 24**, The fire alarm system consists of a thermometer for smoke and a thermometer for heat. The logic gate AND was used to verify the two conditions of high temperature and smoke to a level higher than the threshold limit. A fire alarm is triggered, but if one of the two conditions is met, the alarm is triggered by a red LED.

It is possible to place more than one temperature sensor in different locations of the same station and to show the data represented by a waveform. When an on signal arrives, the alert is done by text and using a red LED. But when there is no fire, there is no signal and the text is no fire. It is possible to show these functions in LabVIEW in **Figure 25**.

4.3 Practical implementation results

In the practical implementation of the project, two main outcomes can be observed. The first output is the hardware of an automated fuel pump station, which interacts with the second outcome, the petrol station monitor system, which is

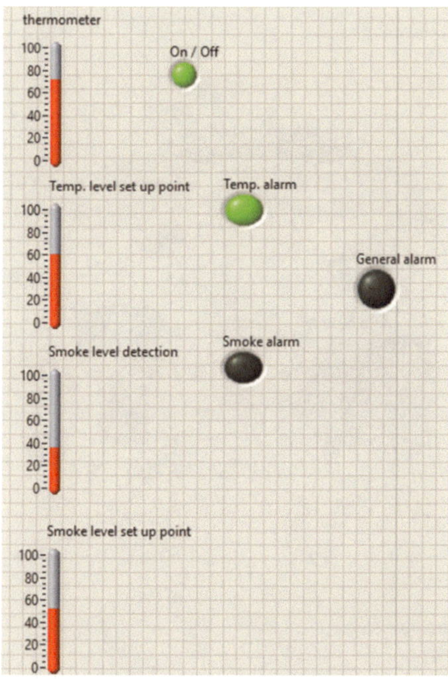

Figure 24.
Alarm fire protection.

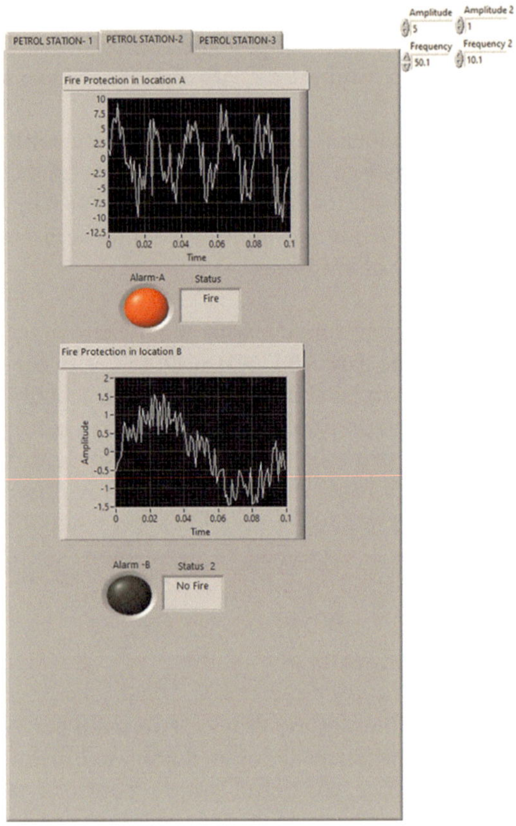

Figure 25.
Alarm fire protection from different locations.

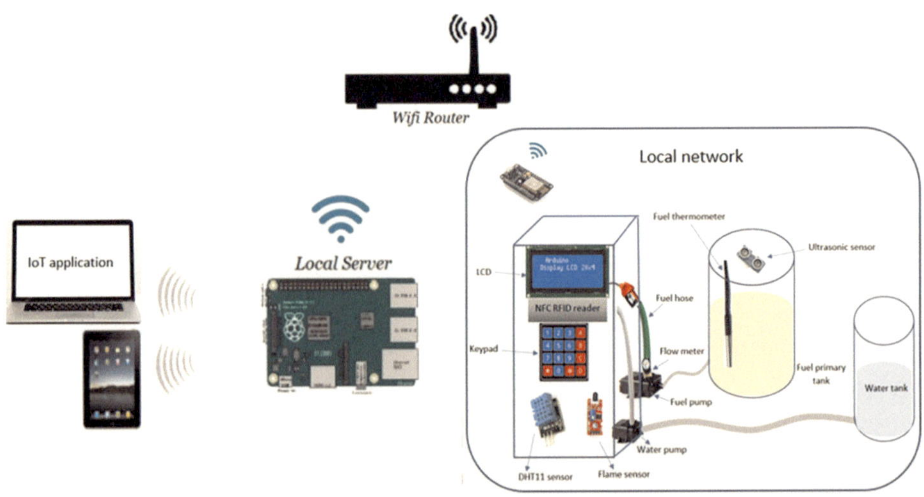

Figure 26.
The proposed model of the hardware system.

represented by the Internet of Things platform (ThingsBoard) installed on a Raspberry Pi as a local server. In this section, the functioning and interface of both are elaborated. **Figure 26** shows the implemented model of the hardware system.

4.3.1 Hardware petrol station system

From the **Figure 27**, it can be seen that the petrol station hardware system is equipped with the customer interface such as LCD screen, NFC RFID reader, temperature and humidity sensor and keypad membrane. The user(manager) is able to interact easily providing this software is user-friendly, and besides that, the hardware is robust, all in one. The nozzles of fuel pump are directly connected.

4.3.1.1 Petrol station hardware—user interface

Initially, when the power board is supplied, the LCD screen shows the functions that the board performs when pressing the letter buttons on the keypad as shown in **Figure 28**.

- When we want to check the balance in RFID tag, we press A letter in keypad as shown in **Figure 29**. After pressing A letter, the LCD screen will display a message telling the customer to insert his card for the purpose of checking as shown in **Figure 30**. After checking the card, a message will appear on the screen showing the user's balance in Iraqi dinar as shown in **Figure 31**.

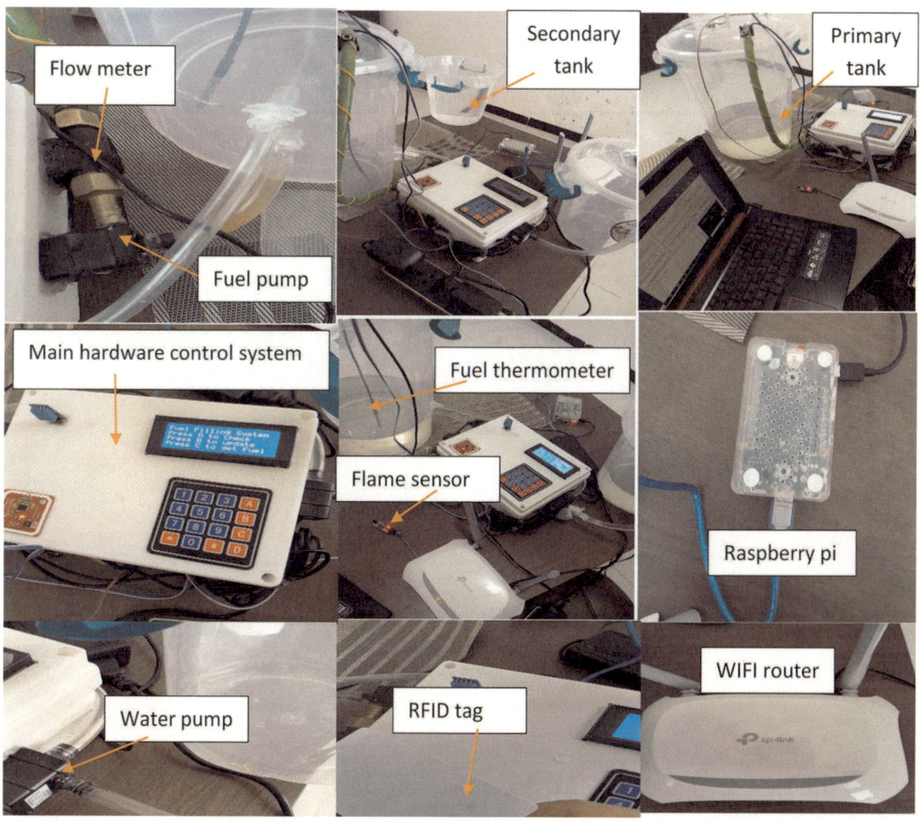

Figure 27.
The overall hardware system.

Figure 28.
Initial display on LCD.

Figure 29.
Press A to check.

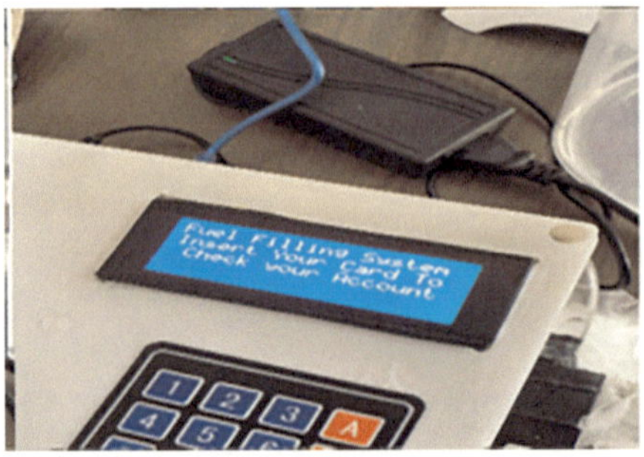

Figure 30.
Insert card to check.

- For the purpose of updating or recharging the balance in the card, we press the letter B as shown in **Figure 32**. After pressing B letter, the LCD screen will display a message telling the customer to insert his card for the purpose of updating as shown in **Figure 33**. After identifying the ID of the card, a message appears on the screen asking the user, the amount to be recharged in the card as shown in

Figure 31.
User ID and your balance in card.

Figure 32.
Press B to update.

Figure 33.
Insert card to update.

Figure 34.
Insert the amount of balance.

Figure 34. After entering the required amount and pressing the letter D, the updated quantity will appear on the screen in addition to the total amount after adding the new amount as shown in **Figure 35**.

Figure 35.
New amount after update.

- In order to obtain fuel, we press the letter C as shown in **Figure 36**. This function is common to all types of regular users (who has a balance and is deducted from his card after filling the fuel) and government users (who does not have a balance on the card, but whose debts are recorded on the government's account). After press C, the LCD screen will display a message telling the customer to insert his card to identify yourself as shown in **Figure 37**. After identifying the ID of the card and thus determining the type of customer, a message appears on the screen telling the customer to enter the amount of fuel to be filled as shown in **Figure 38**. The filling process begins after entering an amount of fuel less than the available balance as shown in **Figure 39**, and the amount deducted from the card and the amount of fuel filled appear on the LCD screen as shown in **Figure 40**. Because the fuel meter has a small error rate, we enter the quantity minus the error percentage.

Figure 36.
Press C to get fuel.

Figure 37.
Insert card to identify.

Figure 38.
Insert quantity of fuel we want to get.

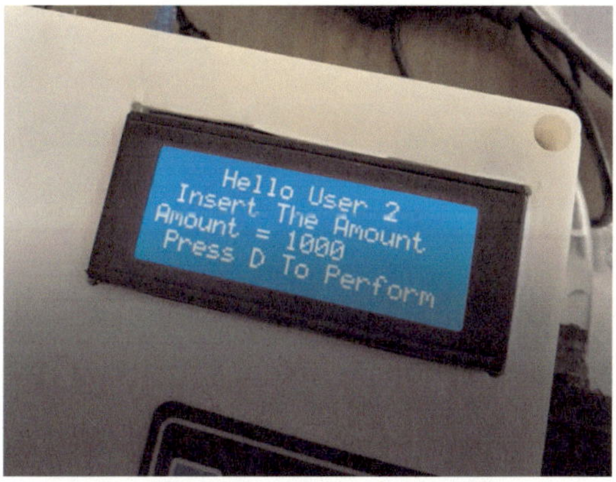

Figure 39.
Insert quantity of fuel we want to get.

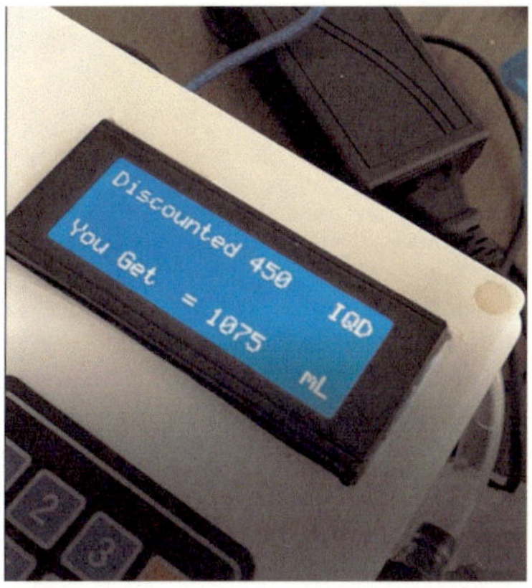

Figure 40.
Price and quantity of filled fuel.

- When entering a quantity of fuel greater than the available balance in the card as shown in **Figure 41**, a message will appear telling the customer that the quantity entered is greater than the available balance as in the **Figure 42**.

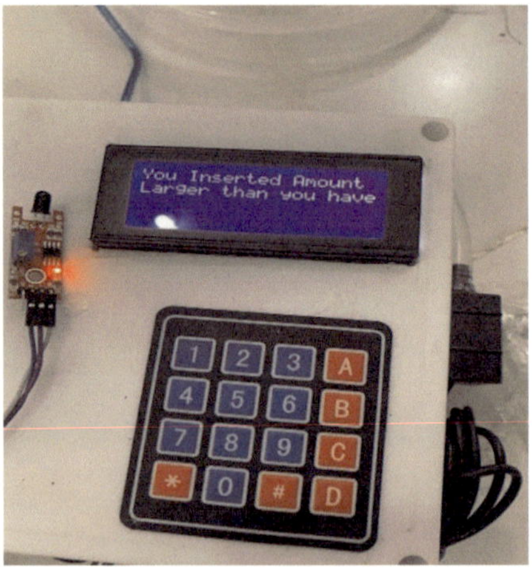

Figure 41.
The quantity entered is greater than the available balance.

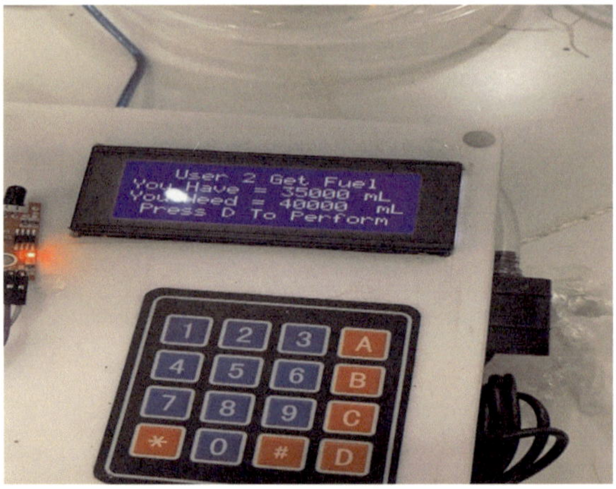

Figure 42.
Entering more fuel than the available balance.

Figure 43.
Check VIP user.

Figure 44.
VIP user identify to get fuel.

Figure 45.
Debts added and total debts.

- When VIP government customer (want to get fuel, initially he swipes his card on the NFC RFID reader to identify as shown in **Figure 43**. Then enter a quantity of fuel VIP wants to get as shown in **Figure 44**. After fuel filling message display on LCD screen telling VIP user about quantity filled and depts as shown in **Figure 45**.

4.3.1.2 Hardware for petrol stations—primary fuel tank

- Level of the primary fuel tank

  By adding Ultrasonic sensor installed at the top of the primary fuel tank for measuring fuel level as in the **Figure 46**. Ultrasonic sensor is installed on the Arduino Mega microcontroller.

- Fuel temperature

  By adding DS18B20 digital thermometer to measure a fuel temperature which enters inside the primary fuel tank as in the **Figure 47**, where by knowing the fuel temperature it is possible to predict the fuel evaporation rate and its density. Fuel thermometer is installed on the ESPS-32 microcontroller.

4.3.1.3 Petrol station hardware- temperature and humidity weather

By adding DHT11 sensor on the petrol station hardware box and connecting it to the ESP-32S microcontroller to represent a second fuel station, that sends temperature and humidity to ThingsBoard server.

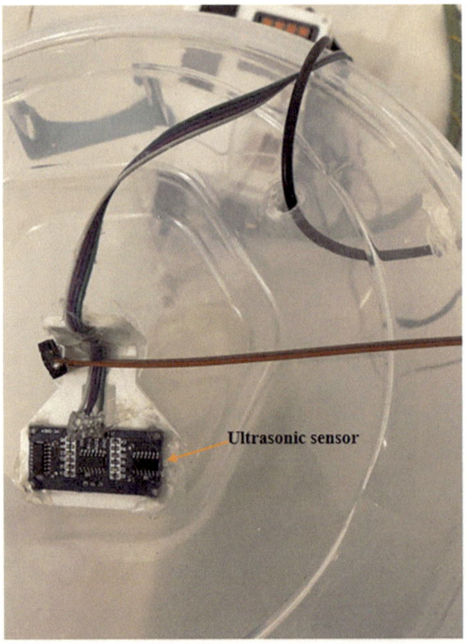

Figure 46.
Ultrasonic sensor at the top of the primary fuel tank.

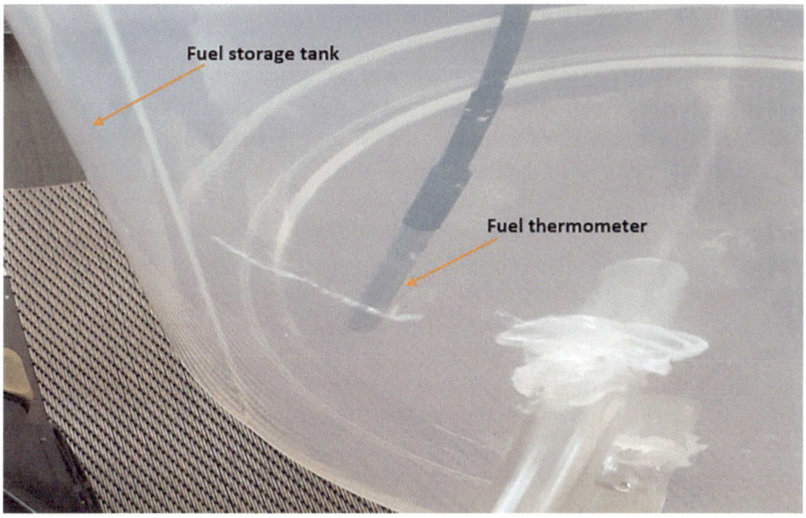

Figure 47.
Fuel thermometer inside the primary fuel tank.

4.3.1.4 Hardware for petrol stations with a fire-protection system

For the purpose of fire protection, a flame sensor was included in this system to sense fires and extinguish them automatically or manually. The water pump was connected to the sides of the main hardware box, which was connected to the flame sensor, which triggers an alarm sound when a flame is detected and sends its data to the microcontroller, which instructs the relay switch associated with the water pump to operate it. This pump can also be controlled manually via a push button and is done Turn on the pump. **Figure 48**, shows the water pump installed on the box and the flame sensor.

Figure 48.
Fire-protection system.

4.3.2 Petrol station monitor system

The monitor system to share data, manage and control is expected to provide the fuel stations manager with information about a petrol station in real time remotely shown in **Figure 49**. The system will show sales and the volume of petrol bought on a specific date, and will provide manager with data as follow:

1. Users interface

2. The primary tank of a fuel

3. Weather conditions (temperature and humidity)

4.3.2.1 User interface for a petrol station monitoring system

All of the customer's actions, such as filling the tank, paying the bill, total sales, the number of cars, debts, and so on, are sent to the server and monitored by the company that follows up on three fuel stations, where each station subscribes to the device's identity and the monitoring company's access token. **Figure 50**, shows the operations performed by the customers as they appear on the dashboard.

4.3.2.2 Petrol station monitor system-primary fuel tank

By monitoring the main fuel tank, it is possible to verify the following:

4.3.2.2.1 Level of fuel

Ultrasonic sensor is installed on the Arduino Mega microcontroller.
The sensor sends fuel level information to microcontroller considering each

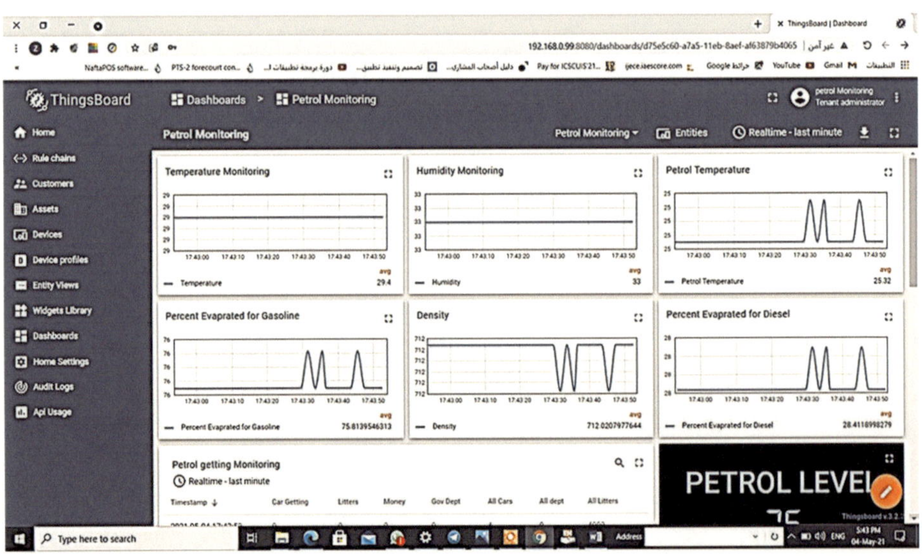

Figure 49.
Petrol station monitoring system Interface.

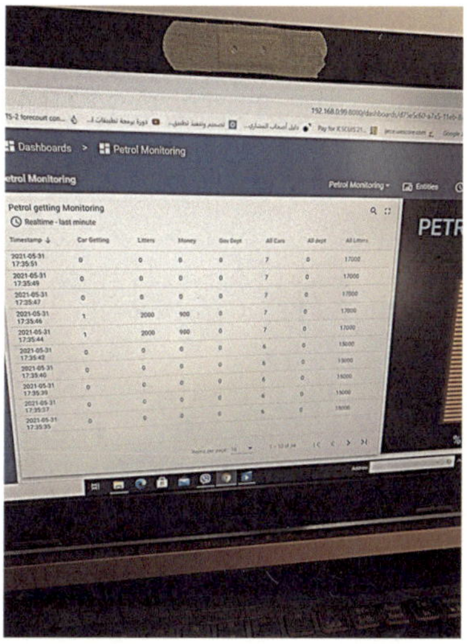

Figure 50.
Petrol monitoring on ThingsBoard dashboard.

microcontroller is a fuel station (MQTT client) and sends its information to the ThingsBoard server as shown in **Figure 51**.

4.3.2.2.2 A fuel temperature

In this thesis project, we use fuel temperature sensor to send their information from microcontroller (as MQTT Clint) to calculate percentage evaporation of fuel

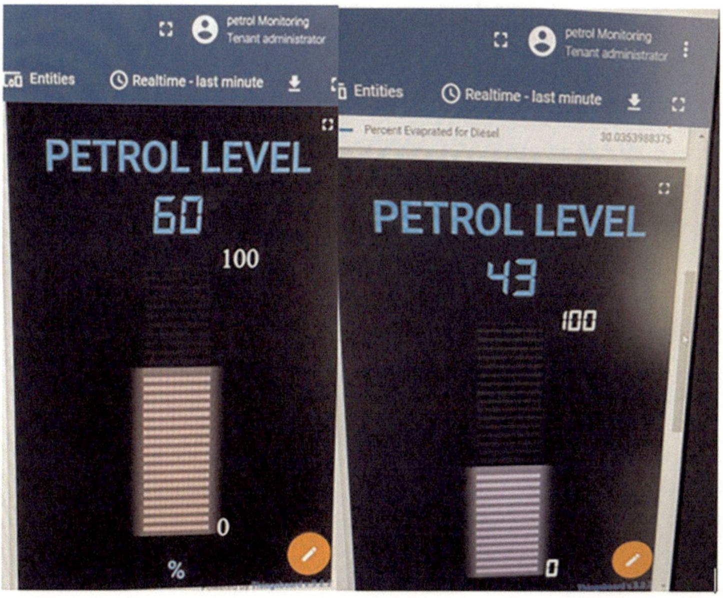

Figure 51.
Monitoring fuel level for two states on the dashboard.

gasoline and diesel and sends its information to the ThingsBoard server (as MQTT broker) installed on Raspberry Pi as a local server or ThingsBoard installed on a cloud server to be shown on the dashboard. **Figure 52a** and **Figure 52b** shows petrol temperature average 25.32 Co and 43.46 Co respectively.

4.3.2.2.2.1 Evaporation rate

Figure 53a and **Figure 53b** shows percentage evaporation of gasoline according to temperature change at average 25.32Co and 34.46 Co respectively, where the effect of temperature appears, with 25.32Co the percentage of evaporation of gasoline becomes 75.8 and then this percentage rises to 83 at 34.46 C°. **Figure 54a** and **Figure 54b** shows percentage evaporation of diesel fuel according to temperature change at average 25.32C° and 34.46 C° respectively, where the effect of temperature appears with 25.32Co the percentage of evaporation of diesel becomes 28.4 and then this percentage rises to 30 at 34.46 Co.

4.3.2.2.2.2 Density

Figure 55a and **Figure 55b** show the variation density of gasoline according to temperature change at average 25.32C° and 34.46 C° respectively, where the effect of

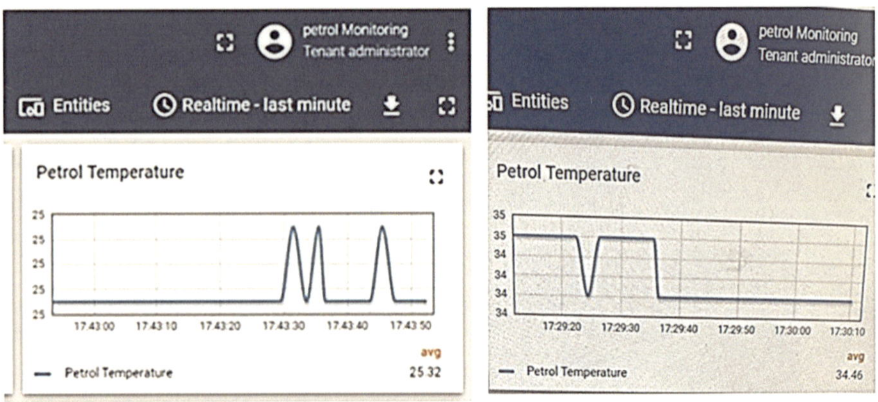

a b

Figure 52.
Gasoline temperature as shown on dashboard.

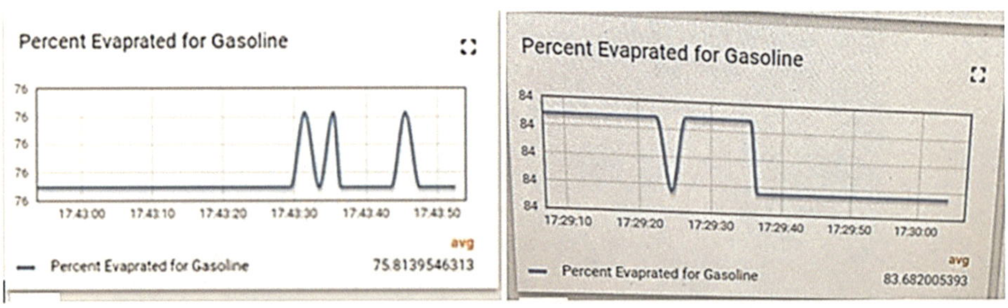

a b

Figure 53.
Percent evaporated for gasoline on dashboard.

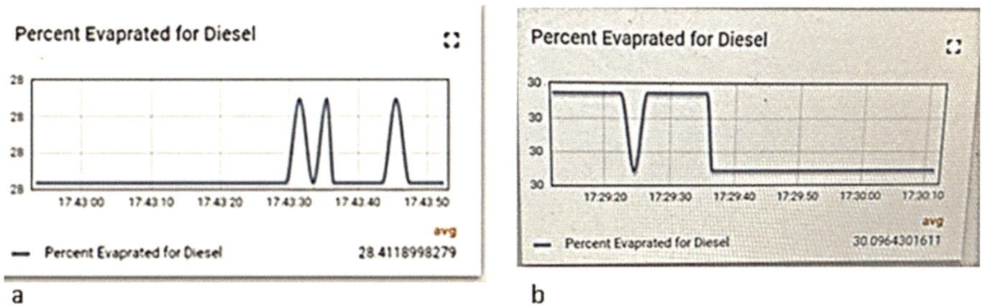

Figure 54.
Percent evaporated for diesel on dashboard.

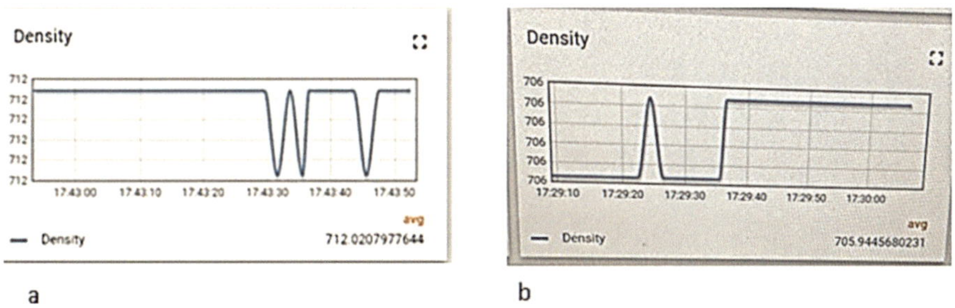

Figure 55.
The density of gasoline vs. temperature on the dashboard.

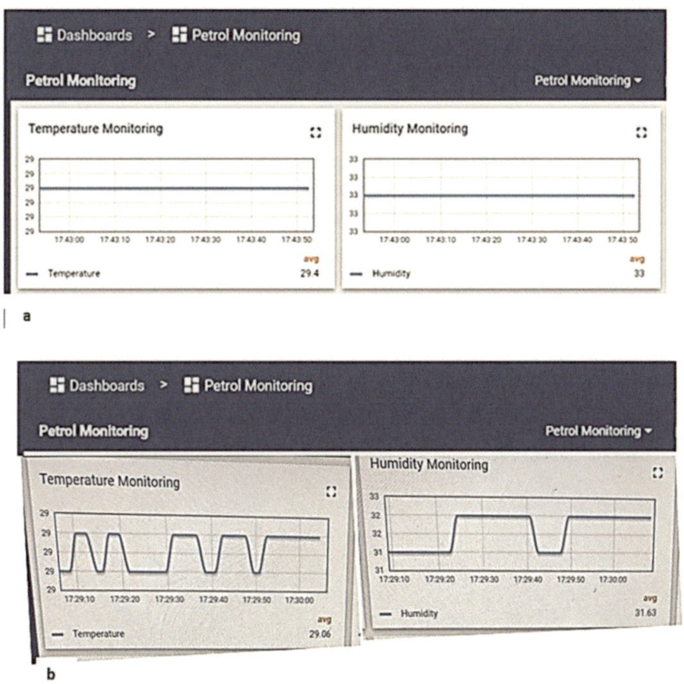

Figure 56.
Temperature and humidity are being monitored on the dashboard for two states from the second fuel station (a and b).

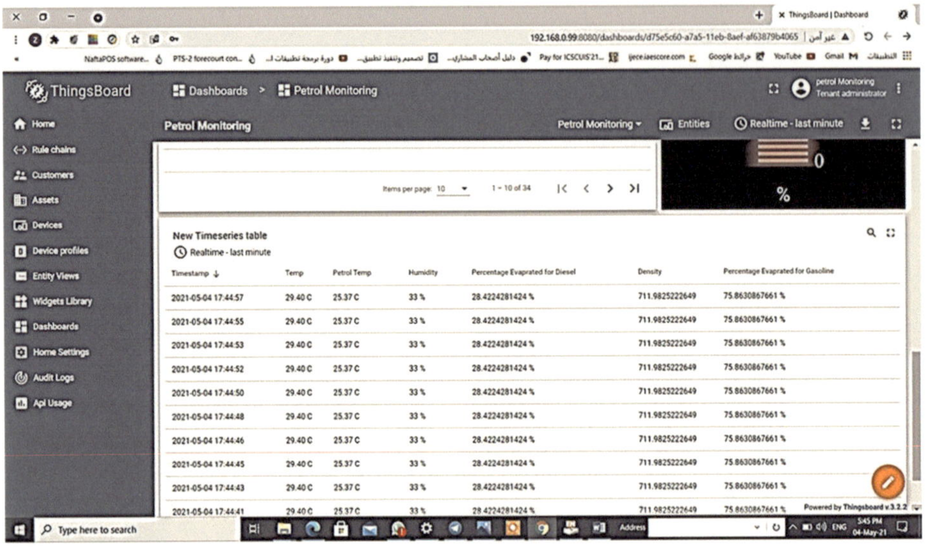

Figure 57.
Timeseries table on the dashboard.

temperature appears with 25.32C° the density becomes 712 kg/m3 and then this density decreases to 7.6 at 34.46 C°.

4.3.2.2.3 Petrol station monitor system-temperature and humidity weather

DHT11 sends temperature and humidity information to a ThingsBoard server installed on a Raspberry and displays this information on the ThingsBoard dashboard, so that the fuel station is monitored and any station that sends its information on the access token and the identity of the device that is connected to the display dashboard. **Figure 56a** and **Figure 56b** shows the temperature and humidity of the second fuel station for two different states as they appear on the dashboard monitoring the three gas stations. Monitoring the temperature of the fuel and its consequences in determining the evaporation rates of fuel (gasoline and diesel fuel), as well as determining the density and the consequent determination of whether the fuel is tainted with invalid mixtures. Monitoring these coordinates from a fuel station whose client MQTT is microcontroller ESP-32S. All of these variables are monitored on the dashboard of the ThingsBoard server as shown in **Figure 57** in timeseries table of the dashboard.

5. Conclusions

The smart petrol station system is expected to solve various problems as stated in the first chapter. This project has added a new technology in which a wireless communication can be established between petrol station stations to monitor and control a system which may be located hundreds of kilometers away. Thus, the monitoring process of the petrol station stations can be simplified by this technology. The implementation and test the automation system which build based on IoT leads to the conclusions below.

1. Connected three petrol station systems to Internet represent the basic structure to IoT of this application in order to share several types of information related to users, environment and database of systems.

2. Practical results show that smart petrol station system-based on IoT scheme is effective and more manageable to resolve the all problems of traditional filling stations.

3. Using the method of distinguishing between government card customers and regular card customers is very useful for reducing corruption.

4. The establishment of this project in retail outlets can reduce the period utilization which associated with the depth of petrol in tanks which was ensured manually with the assistance of big gauging scale, testing the heat and density of petrol, stock totaling, etc.

5. Setting an alarm system in the event of a trip or an overload on the pump, thus extending the life of the fuel pump. Also, Maintenance of this system is very easy and low-cost.

6. Measuring temperature of weather is very useful where there are volume losses in hot weather and volume gains in cold weather of petrol due to thermal expansion.

Author details

Zahra'a M. Baqir[1*] and Hassan J. Motlak[2]

1 Department of Electrical and Electronic Engineering, Thi-Qar University, Nasiriyah, Iraq

2 Department of Electrical and Electronic Engineering, Babylon University, Babylon, Iraq

*Address all correspondence to: zmbhmaly@gmail.com

References

[1] Gupta P, Patodiya S, Singh D, Chhabra J, Shukla A. IoT based smart petrol pump. In: 2016 Fourth International Conference on Parallel Distributed and Grid Computing, Waknaghat, India. Vol. 2016. IEEE; 2016. pp. 28-32

[2] Gowri Shankar EMR, Baskaran D. Automation of petrol bunk using biotelemetry system and petro card. International Journal of Engineering Research & Technology. 2016;**4**:1-4

[3] Wavekar Asrar A, Patel Tosif N, Pathan Saddam I, Pawar H. RFID based automated petrol pump. International Journal for Scientific Research & Development. 2016;**4**:1333-1334

[4] Kumaresan P, Babu Sundaresan Y. IoT based retail automation of fuel station and alert system. In: Materials Science and Engineering Conference Series. IOP Conference Series; 2017. 042072

[5] Chandana K, Chirag M, Burugupalli PKK, Akarsh J. Raspberry pi based RFID smart card refuelling system. In: Perspectives in Communication, Embedded-Systems and Signal-Processing-PiCES Journal, Bangalore, India. Vol. 2. 2018. pp. 149-152

[6] Dongarsane C, Dalavi P, Golandag S, Powar S. Self-operated petrol pump. International Journal of Advance Research, Ideas and Innovations in Technology. 2017;**3**:231-234

[7] BK R. IOT based petrol bunk management for self-operation using RFID and raspberry Pi. International Journal of Recent Technology and Engineering (IJRTE). 2019;**8**

[8] Nitha C, Velayudhan RKR, Rashida MH, Risvana MP, Sreemol CV. Automatic fuel filling system. International Journal of Advanced Research in Computer and Communication Engineering. 2019;**8**: 128-133

[9] Rao SS, Prasad VS. Centralized automation of petrol bunk management and safety using RFID and GSM technology. In: 2017 International Conference on Intelligent Computing and Control (I2C2). Coimbatore, India: IEEE; 2017. pp. 1-5

[10] Makdisie C, Haidar B. Using SCADA & microcontroller for monitoring and automation of petroleum station tanks. International Journal of Engineering Research and General Science. 2017

[11] Deepa D, Deepashree B, Deepu N, Divya R. Multipurpose self fuel dispensing automated framework utilizing RFID prepaid cards. In: 2018 International Conference on Design Innovations for 3Cs Compute Communicate Control (ICDI3C), Bangalore, India. IEEE; 2018. pp. 69-74

[12] Pranto MB, Rahman MM. Vehicle fuel monitoring and management using RFID authentication and telematics notification: An automated petrol refueling & notifying system. In: 2019 International Conference on Advanced Computer Science and Information Systems (ICACSIS), Bali, Indonesia. IEEE; 2019. pp. 477-482

[13] Sembodo BP, Atmajaya H. Design of smart fuel station and management system based on Arduino and internet of things. BEST: Journal of Applied Electrical, Science, & Technology. 2021; **3**:5-10

[14] Gupta N, Brandt SR, Wagle B, Wu N, Kheirkhahan A, Diehl P, et al. Deploying a task-based runtime system on raspberry pi clusters. In: 2020 IEEE/ACM 5th International Workshop on Extreme Scale Programming Models and Middleware (ESPM2). GA, USA: IEEE; 2020. pp. 11-20

[15] Arduino. Arduino Software. (n.d.). Available from: https://www.arduino.cc/en/software

[16] Labcenter Electronics. Proteus Software. (n.d.). Available from: https://www.labcenter.com/

Chapter 7

Hybrid Architectures to Improve Coverage in Remote Areas and Incorporate Long-Range LPWAN Multi-Hop IoT Strategies

Francisco A. Delgado Rajó and Ione Adexe Alvarado Ramírez

Abstract

At the height of M2M communications, there are many alternatives and architectures that present solutions for each case and each environment. The interoperability strategy or the combination of different solutions with adequate flexibility can be solutions to maintain a capacity for easy incorporation of new sensor nodes depending on the coverage or not of operators. In addition to interoperability strategies, this chapter presents some alternatives that include multi-hop techniques, combining different technologies. Special emphasis will be placed on low-power wide area networks systems (LoRa, Narrow Band IoT, LTE, etc.) applied in remote environments, such as nature reserves and ocean or fluvial ecosystems. An estate of art of these areas will be presented, as well as results of different development of our group.

Keywords: wireless sensors networks, IoT, LPWAN, multi-routing, interoperability, IoT protocols

1. Introduction

The number of sensors/actuators implemented in the real world today in various kinds of applications (agriculture, weather observation, marine observation, monitoring of seismic movements, medicine, or industrial applications, for example) is immense. In the world of Internet of Things, almost all of these sensor's nodes have the same requirements in terms of the need for low power, self-configurability, low cost, moderate data rate, and wireless communication capability. Traditionally, the wireless networks used for communication between the nodes and the gateway that connects the things networks with the IP network was of the star type. However, at present, the rapid mass introduction of new elements in networks of this type (Massive Machine to Machine Networks) demands greater flexibility in the topologies and architectures used, as well as the ability to interoperability between the various existing technologies in a transparent way to the end users of the system. In addition, this interoperability allows to extend the coverage of the network if the appropriate protocols and architectures are used that allow, in turn, the adaptation to the conditions of each medium.

In urban and suburban areas, it is easier to resort to cloud connection techniques such as NB-IoT or LTE-M, where 4G or 5G coverage is required to reach the final network, meeting the requirements of data rate, consumption, and low cost. However, many of the possible applications mentioned above are developed in environments, where internet connectivity is very limited or nonexistent, but it is necessary to cover large areas where the sensors are located. These can be large agricultural, livestock farms, or marine or river environments. It is in these cases where hybrid strategies and architectures that combine different technologies and require adequate interoperability between different technologies acquire their importance. In short, to achieve greater coverage there are two main strategies: Development of multi-hop topologies using the appropriate routing protocols of mobile nodes in each case, and on the other hand, the combination of different wireless communication technologies depending on the characteristics of the environment. In the case of remote areas, where there is no cellular network coverage, the immediate thing is to think about LPWAN technologies that use unlicensed bands (LoRa, LoRaWAN, Sixfox, and Weightless). Typically, these techniques employ star technology with a high range. In the case of LoRaWAN, it uses the physical layer of LoRa, based on Chirp spread spectrum, integrating, in addition, the network layer. Where it is necessary to increase coverage, this technology could be implemented by developing clusters of clusters around a gateway that intercommunicates with another until reaching the final destination [1, 2]. This chapter will emphasize networks that allow the rapid incorporation of new mobile nodes, so this structure does not seem the most appropriate in principle. The following are the two strategies mentioned above (Multi-hop and Combination of technologies).

2. Multi-hop strategies

Using the physical layer of LoRa, the payload of the data packet can be used to implement different protocols used by mobile wireless sensor networks (WMSNs). This allows the approach of tree or mesh topologies using these protocols. Keep in mind that, in the case of this chapter, mobile nodes can act as repeaters and end nodes in all cases and that they can appear and disappear from the network at any time. For this reason, the most appropriate protocols are the proactive ones that require the routing table to be constantly updated. Another important dilemma, in these cases, is the parameter used to choose the appropriate route to the destination. On the one hand, minimizing the Time on Air (ToA) of the data packets may be a priority to minimize system latency, or depending on the type of data, the rate of packets received can be used. As is known, in the case of LoRa, several communication parameters can be varied such as the bit rate (BR), the spreading factor (SF), or the transmitted power (Pt). The main challenge is that the system adjusts to the requirements of latency, reliability, or QoS minimizing the energy consumption of the different nodes. For this, there are several works developed on different technologies.

Figure 1 shows an example case of the compromise between the different priorities in a multi-hop topology in the case of LoRa technology. If the end node has direct access to the gateway in both cases if the RSSI route is taken as a parameter to choose the route, in case a) a better response will be obtained, and that route would

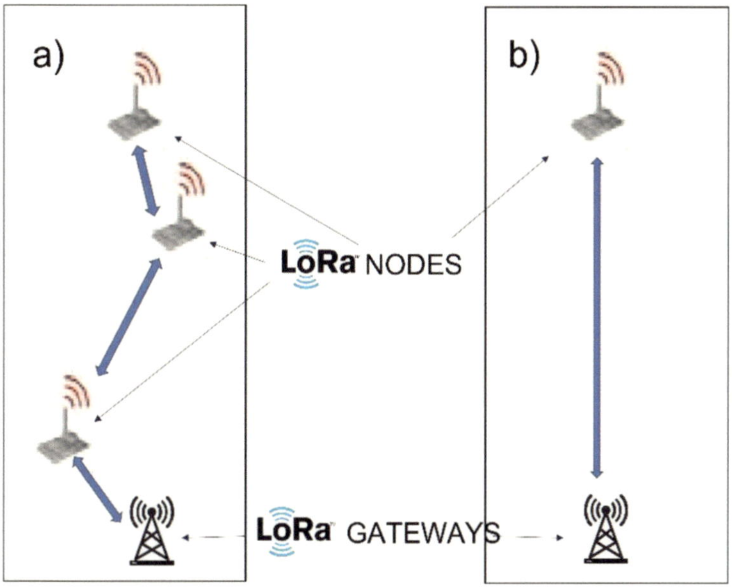

Figure 1.
Two communication cases: a) using multi-hop, b) direct communication with the gateway.

be chosen. The consequence is that it is not necessary to increase the transmission power or the SF achieving a better BR. However, the total ToA would be higher due to the processing of the intermediate nodes, and in addition, they are forced to consume more energy. In case b), it would probably be necessary to increase the transmit power or increase the SF to obtain an adequate Packet Error Rate (PER), although the TOF of the packets would be lower. It is, therefore, important to determine the priority of the parameters to be minimized when choosing the appropriate route depending on the type of data packets, the required latency, or the need to reduce consumption. For this reason, the knowledge of the nature of the entire network is needed and not just find the nearest neighbor. This knowledge must be constantly updated due to the presence of mobile nodes, which impose proactive protocols [3]. In [4], an assessment of the energy efficiency of the network is highlighted according to the network topology and the number of hops. For example, for a given LoRa transceiver, measurements of the energy consumed in a meshed network, such as the one shown in **Figure 2**, are obtained as a function of the range, the number of hops, and the distance between the repeater nodes. Where D is the end-to-end distance (i.e., between the sender and sink node), and d is the distance between each intermediate node.

2.1 Routing protocols

There are multiple routing protocols for the different mobile node network topologies deployed [5]. When selecting one of them for the case of LoRa technology, it is important to consider whether the rapid discovery of new routes or the number of total nodes in the network is critical. In the case of reactive protocols, such as Ad-hoc On-Demand Distance Vector Routing [6], the network is not saturated with route update packets, although in the case of the incorporation of mobile nodes, the use of

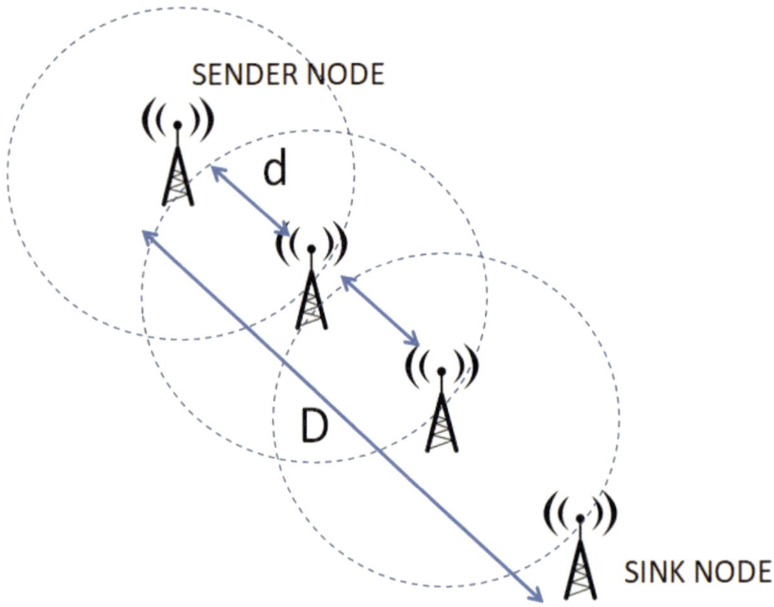

Figure 2.
Mesh topology [4].

these protocols may not be the most appropriate, depending on the time they remain in the network. Others are based on clustering of nodes within the total network (clustering) for large networks, as is the case of LEACH [7], where all nodes can become the central node of a cluster with equal probability. It is a proactive protocol that can be mono-hop or multi-hop [8], both based on extending the life of the network. In the case of the mobile sink-based routing protocol (MSRP) [9], the life of the network is extended avoiding the effect that the nodes closest to the sink of each cluster transmit a greater number of data packets than those that are farther away. For this, the sink nodes become mobile nodes that collect information from several clusters of the total network.

A mixed solution for these mobile node networks, also based on node zone grouping, is the zone routing protocol (ZRP) [10], which groups nodes according to the number of hops required. It is based on the coexistence of two protocols: one intra-zone of proactive type that allows the constant search of the neighboring nodes and another inter-zone of reactive type, thus avoiding the saturation of the network.

For the case of LoRaWAN systems, also based on grouping LoRaWAN nodes around a cluster gateway is the proposal in [11]. This paper proposes a star of stars architecture, where devices or end nodes are connected to one or more gateways using the physical layer of LoRa. Each end device transmitted data to the cluster gateway, and the cluster gateway concatenated data from all the devices in an array and transmitted it to the central gateway. The authors demonstrate that a better energy efficiency is achieved in this star of stars topology.

There are many other applications such as those of large farms, the monitoring of seismic movements, or networks of marine buoys with sensors, where the mobility of the nodes is reduced, although there is still the fact that there are nodes that fall due to lack of battery or new ones are incorporated. This chapter focuses primarily on these cases.

Optimized Link State Routing protocol (OLSR) [12] is a proactive protocol that adapts link state routing for use in mobile ad-hoc networks. It allows multipoint relays. Each node selects a set of its single-hop neighbors to act as relays this information is shared between nodes making this protocol well-suited to large and dense networks with random and sporadic traffic. Destination-sequenced distance vector routing protocol (DSDV) [13] incorporates a sequence number to their routing tables and only refreshes the routing tables when receiving a DSDV packet with a higher sequence number than the node already has. This makes this protocol suitable for fixed node networks or with low-mobility nodes. Another reactive protocol used in these kinds of networks and maybe the most appropriate for the case above explained is Ad-hoc On-Demand Distance Vector Routing (AODV) [14], where each node initializes the routing discovering through a route request packet answered by all network neighbors and during the network recognition process a reverse path is created. Dynamic source routing (DSR) supposes an evolution of this protocol, where a list of hops from source to destination is collected in the RREQ packet as it travels through the network. This list allows implementing a total cost parameter of the route if information about SF, BR, and power required is added. This could be a solution for the problem described in **Figure 1**.

An IoT-oriented protocol is Pv6 routing protocol for low power and Lossy Networks (RPL) proposed by The IETF routing over low-power and Lossy Networks working group [15] RPL]. RPL is a gradient routing technique [16] that considers a WSN as a direct acyclic graph (DAG) rooted at the sink. The goal of RPL is to minimize the costs to reach any sink (from any sensor) by means of an objective function. This function can be defined in many ways adapting it to the operating scenario. Some authors [17] have developed a LoRa network RPL based. In this case, the optimal per-link spreading factor (SF) is one of the RPL objective functions (OF0) in order to compute rank, using the selected LoRa SF as routing metric.

Other approaches try to minimize the energy consumption of the LoRa nodes in a multi-hope network by minimizing the distance to the best neighbor taking into account the node state (busy or free). This is the energy-efficient multi-hop communication solution (e2McH) proposed in [18]. A more complex solution to improve coverage an energy consumption is presented in [19], using variable neighborhood search (VNS) and a minimum-cost spanning tree algorithm employing LoRaWAN end nodes and LoRa nodes as repeaters. In this work, the initial solution approach to find the multi-hop route to the final gateway is based in the PRIM algorithm [20] to find minimal spanning tree, storing the values of SF, BW, and Pt for each node. After that, variable neighborhood search (VNS) is employed to change node characteristics. The authors propose a stochastic algorithm that changes neighborhood structures to find local minimal solutions using a function objective based in the energy per useful bit transmitted. This proposal is designed for a three-level network: Level 1: gateway, level 2: repeaters, and level 3: sensors.

In summary, all these works seek the choice of appropriate protocols to minimize, mainly, the energy consumed, and the Time on Air (ToA) parameters of the packets transmitted along multi-hop networks. Some proposals are based on the location of the nodes, and others are based on the suitability of the configurable parameters of the LoRa nodes (SF, Pt, or BW). In certain use cases, it is also possible to extend the network coverage by combining different wireless communication technologies depending on the environmental conditions. The latter gives greater flexibility to the design of the final network adding more dynamic adaptability to the environment. In the next section, different use cases that develop this type of strategy are presented.

3. Interoperability strategies

There are currently many radio technologies (RAT) available when thinking about the interconnection of sensor networks. Depending on the coverage radius, BR, QoS, or consumption of these devices one or the other can be selected. To obtain greater flexibility in the design of networks that adapt to different scenarios, it may be interesting to combine several of them. The options to achieve interoperability between the different technologies can be very varied, both in a software way or in a hardware way. The IoT network designer is often faced with this dilemma when it comes to choosing one technique or another. Many manufacturers already develop, for example, configurable chipsets that can integrate different radio access technologies (RAT's) [21–23], usually using different transceivers interconnected with a microcontroller unit (MCU) through SPI or UART interfaces. Through these combinations, it is not only intended to expand the coverage of the sensor network, but also to be able to reach areas where a single specific technology cannot do due to environmental conditions (obstacles, indoor/outdoor scenarios, density of nodes, etc.).

Another interesting alternative is the one proposed by [24] in which interoperability between different technologies is achieved through software defined networking (SDN). The objective is the coexistence between different radio access technologies in terms of protocols, coding, or signaling in a transparent way to the final application in the IP network. The architecture is organized into six levels, and the interoperability is classified into: device level, syntactic level, semantic level, network level, and platform level. Each technology has its own modulation, coding scheme, protocols, routing methods, or end-to-end applications communications. The proposed architecture is based on the implementation of software defined radio (SDR), network function virtualization (NFV), and SDN to solve the interoperability problem.

This chapter presents cases of hardware interoperability between two LPWAN technologies or between LPWAN and PAN technologies. A classification could be made into two types of interoperability from the Hardware point of view:

1. Inter-network interoperability: Implementing central nodes that act as a gateway between the two technologies that are separated in the network. In this case, a hierarchical structure of the network would be obtained, and the "translation" point would be only one.

2. Distributed Interoperability: Developing end nodes, routers, or central nodes that implement the two technologies and can choose between one and the other, or even redundant communications using both. This case gives greater flexibility to the network and allows automatic adaptation to the environment of the nodes of the same.

Figure 3 shows the two interoperability models mentioned.

3.1 Inter-network interoperability

This philosophy is applicable, mainly if the conditions of the environment set the radio access technology to be used in different areas of the total network, or to separate the traffic of one network from another implementing a topology contemplating subnets. It is possible, for example, that in cases of long extensions of land or marine or river environments, there are areas, where 4G or 5G coverage is available, such as

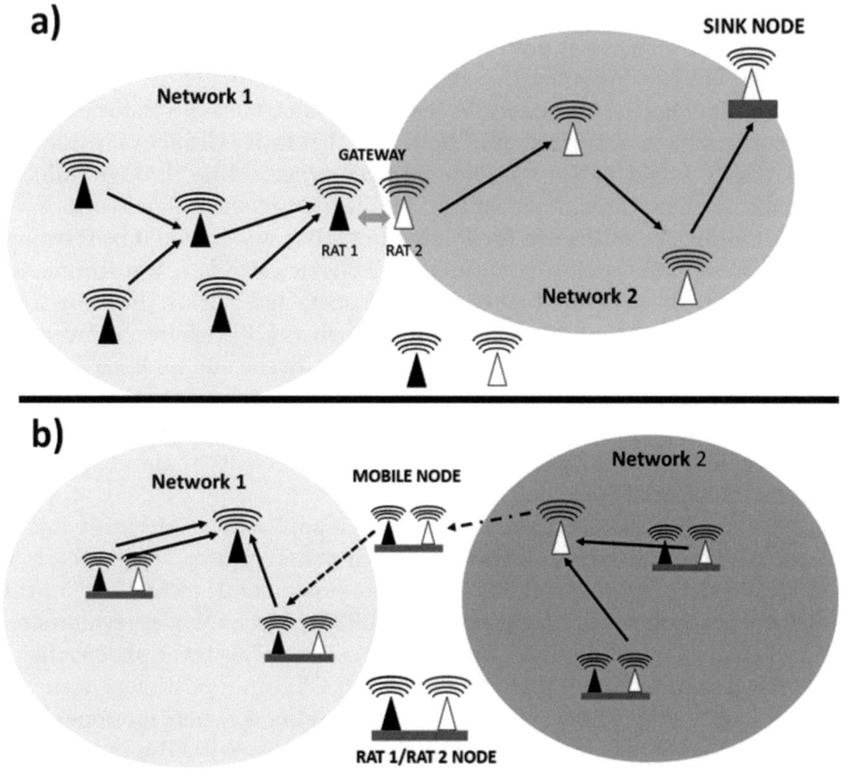

Figure 3.
Two kinds of interoperability possibilities including multi-hop: a) inter-network, b) distributed.

urban or suburban environments and others without this type of network deployed. In the case of long-range technologies, we would choose options such as NB-IoT or LoRa, for example. Another possibility is to interconnect a cellular network already deployed with a network of sensors located at long distances and indoors, as would be the case of greenhouses or fish farms. In these cases, the philosophy used would be to create gateway nodes between the different RAT's. Its function is to adapt protocols, contents of the payloads, and control of access to the medium. In addition, this separation enables the possibility of existence of pre-processing before communication between networks, reducing the traffic load between subnetworks and in the entire network (Edge Computing) [25]. The latency in cases of Sensor/Actuator interaction can also be reduced with this arrangement since the data does not have to be sent to the rest of the network in the event that they are on the same subnet. Examples of this philosophy are presented below.

3.1.1 LoRa-ZigBee interoperability

As seen in previous sections, one of the biggest challenges of this type of network is the reduction of power consumption. ZigBee technology, based on the 802.15.4 [26] standard, is widespread in sensor communications networks. It allows flexible network configuration, moderate ranges (10–100 meters), low-power consumption, and data transmission rates from 40 to 250 Kbps depending on the band of use (915 MHz or 2.4 GHz). It is more susceptible to be used indoors and in the absence of obstacles.

In the comparison between LoRa and ZigBee, the latter is characterized by its lower cost, as well as its lower power consumption and a higher BR, to consist of a shorter range. The combination, therefore, of both technologies can mean energy and economic savings in some cases. In the case of smart buildings, for example, there are performance comparisons of both [27] that ratify the above in addition to the fact that LoRa technology provides greater penetrability through walls or cement walls. In these cases, it would be possible to implement a network that follows the philosophy of **Figure 3a**, where the ZigBee nodes would be forming a subnet within the same enclosure without wall obstacles and are interconnected by LoRa links, longer range and more robust to obstacles. Within the same smart buildings ecosystem, there are other proposals such as [28], where the inverse strategy is presented, that is, the interconnection with the end nodes is formed by LoRa links, while the ZigBee links implement the connection with the central data collector and the IP network. In this way, the BR of the ZigBee technology is used in the final stretch of the network, as well as its greater security thanks to the AES 128 [29] encryption algorithm.

Another environment, where interoperability is applicable, is that of remote natural parks or crops, where greenhouses appear scattered over a large area. In these cases, it is essential to sensorize them. ZigBee is a very suitable technology for this, case, but, if it is necessary to cover large areas, LPWAN technologies result more suitable to implement the backbone of the complete system. This is the philosophy that this research group has followed in [30], adding VLC communications systems [31]. The overall architecture of the system is shown in **Figure 4**, where interoperability between the different technologies (LoRaWAN, ZigBee, and VLC) is implemented in each of the access points of the ZigBee subnets.

In case that there is communication between nodes of the same ZigBee subnet, it is forwarded directly by each access point, avoiding the increase in latency in a possible sensor-actuator communication or in the case of alarm action. When making the gateway between the two RAT's it is necessary to take into account the size of the payloads of both. In this case, the Waspmote [23] hardware platform of Libelium has been used, which contains an ATmega1281 microcontroller with a series of connected sensors and allows two simultaneous communication modules connected *via* SPI bus. The microcontroller is responsible for storing the messages from each node of the network and retransmission by the necessary technology. In the case of ZigBee-LoRaWAN, a fragmentation of the data packets is necessary as shown in **Figure 5**.

In addition, the use of VLC is proposed for the transmission to a mobile device [32] of contents stored in the memory of the access point or transmitted through LoRaWAN from the central gateway of the network. The goal was to create a cellular network, where ZigBee and VLC coverage were given to each cell.

3.1.2 Lora-NB-IoT

Another contribution within the strategies reflected in **Figure 4**. a is based on LoRa and NG-IoT technologies [33] where a network of LoRa-NB-IoT gateways is implemented between areas where there is 4G or 5G coverage, and those where there is not. NB-IoT operates in the licensed bands associated with mobile operators and allows communications with low consumption and with BR from 120 to 160 Kbps. In addition, it has a low latency of the order of 1 to 10 sec. This can be an inconvenience, but for sensor networks, it is not a critical aspect. In this case of hybrid network, the

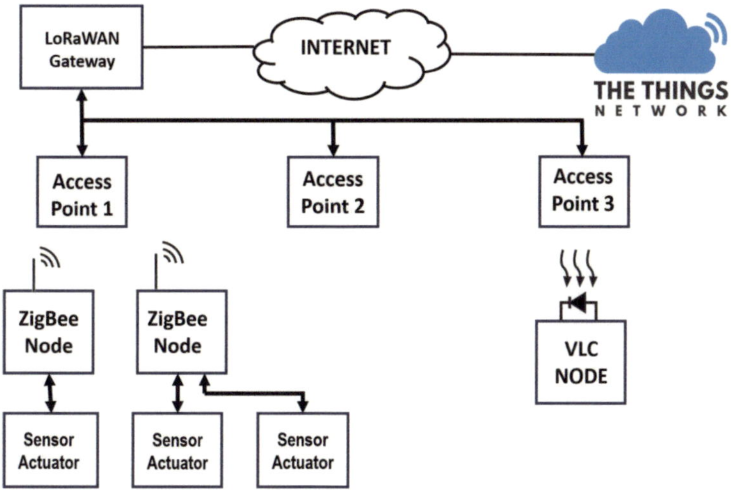

Figure 4.
Global network architecture [30].

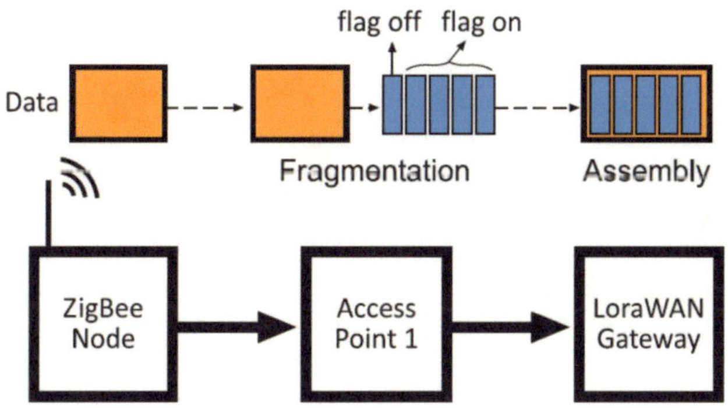

Figure 5.
Fragmentation in a ZigBee-to-LoraWAN communication.

LoRaWAN subnet is responsible for communications that require greater immediacy, and the NB-IoT network forms the gateway to the IP network of all sensor data. The architecture used is based on three layers: sensor nodes layer, forwarding layer, and cloud layer. The end-to-end communication of the network is carried out using the MQTT messaging protocol [34], based on the publish/subscribe philosophy, which is supported by all nodes in the network.

3.2 Distributed interoperability

An example of the philosophy presented in **Figure 3**. b of distributed interoperability is presented in [35], where a multi-RAT architecture is proposed for each node using two LPWA technologies: LoRa and NB-IoT. The latter allows greater BR, in addition to direct connection to IP technology through the mobile network. Therefore, the use of this technology implies the existence of coverage by a mobile operator to be able to implement the network. In the uplink, the use of NB-IoT may require higher power consumption due to the need to implement the entire IP protocol stack. In this case, the use of LoRa may be less energetically costly. The authors propose a series of possible functionalities thanks to these multi-RAT nodes, such as:

- "a security or malfunction detection system with heartbeat status messages over LoRaWAN and emergency alarm traffic over either NB-IoT or both technologies,
- an assisted living wearable with low priority traffic over LoRaWAN and emergency traffic over NB-IoT,
- actuators with heartbeats or status reports over LoRaWAN and direct control loop traffic over NB-IoT,
- a shipment tracking system using NB-IoT whenever available and private LoRaWAN in remote areas (in warehouses, on ships in the sea, etc.)."

Precisely on this last point is based the use case proposed in the next section of this chapter.

4. Proposed use case

In coastal maritime environments, there is a need to monitor parameters such as currents, salinity, winds, temperature, water oxygenation, etc. The sensors responsible for these measurements can be assembled in buoys or, in some cases, in ships that usually sail on certain routes in the area. These ships can act as mobile nodes of the network collecting information from their own sensors or those of fixed buoys at a point and act as repeaters. An appropriate LPWAN technology for the interconnection of these nodes to the gateway could be LoRa, which also has great advantages due to its characteristics in this environment. There are multiple studies analyzing the propagation of radium in this ecosystem [36, 37].

On the other hand, in many archipelagos, there are certain areas in which there is mobile network coverage since operators provide it on routes transited by habitual maritime traffic, such as car and passenger ferries. These vessels are likely to be used as mobile nodes that would act as a gateway between technology, such as LoRa and NB-IoT. The use of nodes that combine these two technologies as proposed in the previous section is very useful giving greater versatility to the network and choosing one or the other always depending on the conditions. Not only offering a path to the cloud through land but also using the vessels themselves as mobile sink nodes agglutinating a cluster of sensor nodes in their route. **Figure 6** shows this use case.

The implementation of hybrid LoRa/NB-IoT nodes allows the right technology to be chosen in each case to reach the final application in the cloud. The nodes have a memory that stores data from sensors that could not be sent during times of absence of connection. In this particular case, non-proactive protocols could be contemplated because the appearance and disappearance of nodes in the network are not so fast, so the updating of routes does not have to be so continuous and the number of nodes in the network at each moment is not as large as in urban or suburban environments.

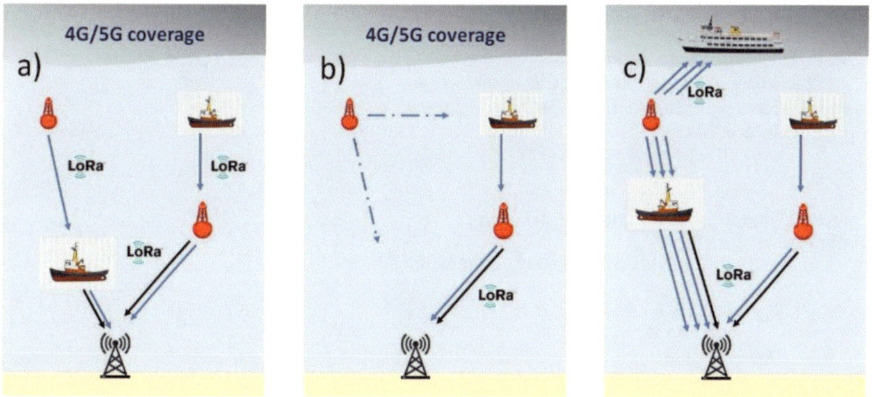

Figure 6.
Proposed use case. a) LoRa multi-hop network with a central gateway, b) connection loss, and c) recovered path and retransmission of the saved data. Incorporation of mobile sink node with 4G/5G connection and interoperability.

DATA	Flags	Hop Count	Destination Address	Source Address	Seq. Number	Sensors Data	Power Level

Figure 7.
Main frame structure.

Destination Address	Sequence number	Hop Count	Next Hop	Pointer to a list of prec.

Figure 8.
Routing table entry.

The protocol used by our group in this case, for the multi-hop LoRa network is based on the Ad-hoc On-Demand Distance Vector (AODV), with some small variations that allow the forwarding of sensor data accumulated in the queues implemented by the nodes. In addition, a field appears in the routing tables of each node that contains a pointer to a list of precursors to store the path of the packet to reach the node in question. So far, its operation has been proven in different types of links (aquatic and suburban). It is currently at the stage of action. **Figure 7** shows the format of the data package used, and **Figure 8** shows the routing table entry.

Figure 9 shows the pseudocode of the main algorithm responsible for finding a route to the gateway directly or through another node to transmit sensor data and the level of power used. That allows the measurements to be carried out in the end.

Simulations have been carried out with Matlab using the Longley-Rice model. This has been done to see the behavior in the marine environment from real points between the gateway on land and an isolated node. **Figure 10** shows the results obtained on a sea route between the islands of Gran Canaria and Tenerife.

The system has been implemented using Libelium Waspmote with a LoRa communications module based on Semantech's SX1272. Depending on the manufacturer and depending on the SF and BW, it has different operating modes. The worst sensitivity for the lower SF case is −114 dBm and − 134 dBm for the maximum SF (12).

```
Algorithm 2: sendSensorsInfo function

Result: Creates a sensor info message and sends it directly if a route to LoRa Gateway (LGW)
available or looks for a route to LGW in other case.
Input:  destiny_address = LoRa Gateway Address;
        flags = 101;
        timeout = timeout specified in loop function


route = searchTableEntry(destiny_address);
if(route == NULL) then
        route = discoverRoute(destiny_address, flags);

        if(route == NULL then
                return -1;
        end

end

powerLevel = route.powerLevel;

createPayload();

for(i = 0; i < MAX_DATA_RETRIES; i++) do
        r = sendMsg(route.nextHop, payload, payloadLength, DATA, timeout);

        if(r == 0) then
                return 0;
        else
                increasePower();
        end
end

processRERR();

deleteTableEntry();

return -1;
```

Figure 9.
Main algorithm pseudocode.

Figure 11 shows the RSSI measurements in a coastal environment with a gateway and in different locations of a mixed coastal environment.

5. Conclusions

This chapter has carried out a review of the options currently available to improve coverage in sensor networks, especially those covering large areas in remote areas (LPWAN). A classification of two strategies has been proposed to obtain greater coverage and flexibility based on interoperability between existing technologies. In addition to the state of the art, a contribution made by our research group has been presented in which multi-hop and interoperability strategies are combined by adding new techniques, such as VLC and OCC. Finally, a use case is presented in which all the techniques seen can be accommodated helping to solve a specific particular problem using a low-cost solution. This prototype is in the measurement period, although the results obtained show good performance in terms of TOA and latency. A very flexible and adaptive architecture is achieved that avoids the loss of data due to the fall of a route or a node. As a future line, a node with NB-IoT / LoRa connectivity is incorporated into the cellular network.

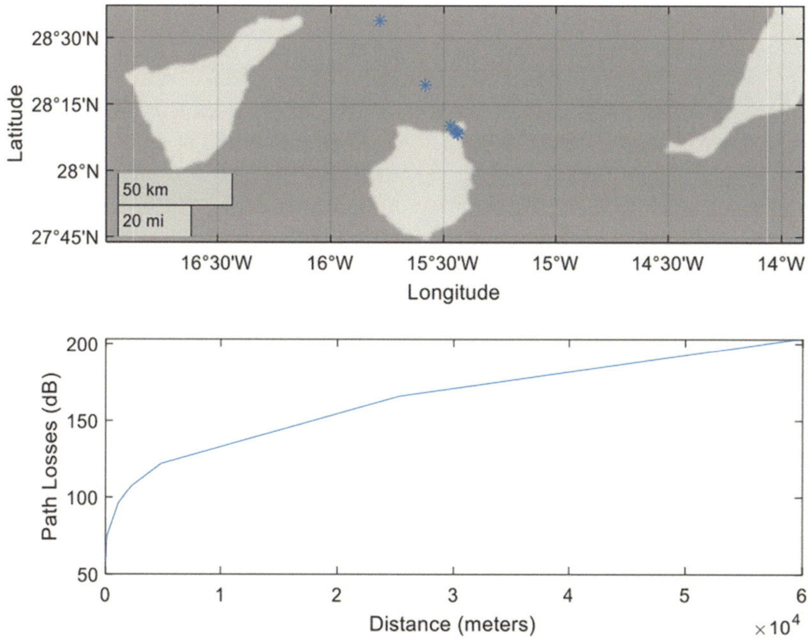

Figure 10.
Path losses versus distance.

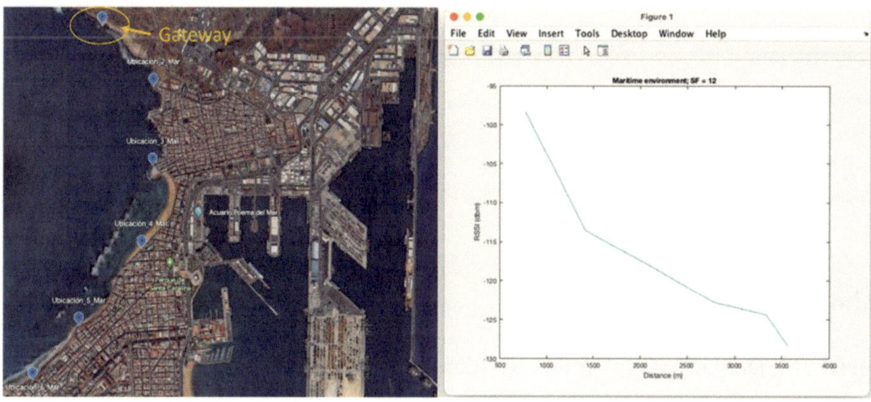

Figure 11.
Measured RSSI versus distance.

Conflict of interest

The authors declare no conflict of interest.

Author details

Francisco A. Delgado Rajó[1*] and Ione Adexe Alvarado Ramírez[2]

1 Instituto para el Desarrollo Tecnológico y la Innovación en Comunicaciones (IDeTIC), Universidad de Las Palmas de Gran Canaria (ULPGC), Las Palmas, Spain

2 School of Telecommunications and Electronics Engineering, Universidad de Las Palmas de Gran Canaria, Spain

*Address all correspondence to: paco.rajo@ulpgc.es

References

[1] Akyildiz IF, Su W, Sankarasubramaniam Y, Cayirci E. A survey on sensor networks. IEEE Communications Magazine. 2002;**40**(8):102-114

[2] Chelouah L, Semchedine F, Bouallouche-Medjkoune L. Localization protocols for mobile wireless sensor networks: A survey. Computers and Electrical Engineering. 2017:1-19. DOI: 10.1016/j.compeleceng.2017.03.024

[3] Walikar G, Biradar R. A survey on hybrid routing mechanisms in mobile ad hoc networks. Journal of Network and Computer Application. 2017;**77**:48-63

[4] Ochoa M, Guizar A, Maman M, Duda A. Evaluating LoRa energy efficiency for adaptive networks: From star to mesh topologies. In: 2017 IEEE 13th International Conference on Wireless and Mobile Computing, Networking and Communications (WiMob). Rome, Italy; 2017. pp. 1-8. DOI: 10.1109/WiMOB.2017.8115793

[5] Alotaibi E, Mukherjee B. Survey paper: A survey on routing algorithms for wireless ad-hoc and mesh networks. Computer Networks. 2012;**56**:940-965

[6] Perkins C, Belding-Royer E, Das S. Rfc 3561-ad hoc On-Demand Distance Vector (AODV) Routing. 2003 [online]. Available from: https://tools.ietf.org/html/rfc3561

[7] Nayebi A, Sarbazi-Azad H. Performance modeling of the LEACH protocol for mobile wireless sensor networks. Journal of Parallel and Distributed Computing. 2011;**71**(6):812-821. DOI: 10.1016/j.jpdc.2011.02.004

[8] Mezghani O, Abdellaoui M. Improving network lifetime with mobile LEACH protocol for wireless sensors network. In: Proceedings of the 15th International Conference on Sciences and Techniques of Automatic Control & Computer Engineering-STA'2014; December 21-23; Hammamet. Tunisia: IEEE; 2014. pp. 613-619

[9] Nazir B, Hasbullah H. Mobile sink based routing protocol (MSRP) for prolonging network lifetime in clustered wireless sensor network. In: Proceeding of the 2010 International Conference on Computer Applications and Industrial Electronics (ICCAIE 2010); 5-7 December 2010; Kuala Lumpur. Malaysia: IEEE; 2010. pp. 624-629. DOI: 10.1109/ ICCAIE.2010.5735010

[10] Beijar N. Zone Routing Protocol (ZRP). Networking laboratory. Helsinki University of Technology, Finland. 2002;**9**(1):12

[11] Aslam MS, Khan A, Atif A, Hassan S, Mahmood A, Qureshi H, et al. Exploring multi-hop LoRa for green smart cities. IEEE Network. 2019;**34**(2):225-231 DOI: 10.1109/MNET.001.1900269

[12] Clausen T, Jacquet P. Rfc 3626: Optimized Link State Routing Protocol (OLSR). 2003. Available from: https://www.rfc-editor.org/rfc/rfc3626.txt [Accessed: April 25, 2017]

[13] Perkins CE, Bhagwat P. Highly dynamic destination-sequenced distance-vector routing (DSDV) for mobile computers. ACM SIGCOMM Computer Communication Review. 1994;**24**(4):234-244

[14] Perkins C, Belding-Royer E, Das S. Rfc 3561 - ad hoc on demand distance vector (AODV) routing. 2003. Available from: https://tools.ietf.org/html/rfc3561 [Accessed: April 25, 2017]

[15] Winter T, Thubert P. RPL: IPv6 routing protocol for low power and lossy networks. Draft-ietf-roll-rpl-15 (work in progress) IETF ROLL working group. 2010. Available from: https://www.rfc-editor.org/info/rfc6550

[16] Watteyne T, Pister K, Barthel D, Dohler M, Augé-Blum I. Implementation of gradient routing in wireless sensor networks. IEEE Global Telecommunications Conference (GLOBECOM). 2010:1-6. DOI: 10.1109/GLOCOM.2009.5425543

[17] Sartori B, Thielemans S, Bezunartea M, Braeken A, Steenhaut K. Enabling RPL multihop communications based on LoRa. In: 2017 IEEE 13th International Conference on Wireless and Mobile Computing, Networking and Communications (WiMob). Rome, Italy: IEEE; 2017. pp. 1-8. DOI: 10.1109/WiMOB.2017.8115756

[18] Anedda M, Desogus C, Murroni M, Giusto DD, Muntean G-M. An energy-efficient solution for multi-hop communications in low power wide area networks. In: 2018 IEEE International Symposium on Broadband Multimedia Systems and Broadcasting (BMSB). Valencia, Spain: IEEE; 2018. pp. 1-5. DOI: 10.1109/BMSB.2018.8436722

[19] Gava MA, Rocha HRO, Faber MJ, Segatto MEV, Wörtche H, Silva JAL. Optimizing resources and increasing the coverage of internet-of-things (IoT) networks: An approach based on LoRaWAN. Sensors. 2023;**23**:1239. DOI: 10.3390/s23031239

[20] Moret BM, Shapiro HD. An empirical analysis of algorithms for constructing a minimum spanning tree. In: Proceedings of the Workshop on Algorithms and Data Structures; 14-16 August 1991; Ottawa, ON, Canada; 1991. pp. 400-411. DOI: 10.1007/BFb0028279

[21] Texas Instruments. CC1350 SimpleLink Ultra-Low-Power Dual-Band Wireless MCU, SWRS183A. 2016. Available from: https://www.ti.com/lit/ds/symlink/cc1350.pdf?ts=1698927098423&ref_url=https%253A%252F%252Fwww.google.com%252F

[22] Nordic Semiconductor. nRF52840 Objective Product Specification v0.5.1. July 2017. Available from: https://infocenter.nordicsemi.com/pdf/nRF52840_OPS_v0.5.1.pdf

[23] Libelium Waspmote v15. https://development.libelium.com/waspmote-technical-guide/

[24] Arzo ST, Zambotto F, Granelli F, Bassoli R, Devetsikiotis M, Fitzek FHP. A translator as virtual network function for network level interoperability of different IoT technologies. In: 2021 IEEE 7th International Conference on Network Softwarization (NetSoft). Tokyo, Japan: IEEE; 2021. pp. 416-422. DOI: 10.1109/NetSoft51509.2021.9492677

[25] Khan WZ, Ahmed E, Hakak S, Yaqoob I, Ahmed A. Edge computing: A survey. Future Generation Computer Systems. 2019;**97**:219-235. DOI: 10.1016/j.future.2019.02.050

[26] IEEE 802.15 WPAN™ Task Group 4. http://www.ieee802.org/15/pub/TG4.html

[27] Liu Z, Li Y, Zhao L, Liang R, Wang P. Comparative evaluation of the performance of ZigBee and LoRa wireless networks in building environment. Electronics. 2022;**11**:3560. DOI: 10.3390/electronics11213560

[28] Ali A, Zorlu Partal S. Development and performance analysis of a ZigBee and LoRa-based smart building sensor network. Frontiers in Energy Research. 2022;**10**. DOI: 10.3389/fenrg.2022.933743

[29] National Institute of Standards and Technology (NIST). FIPS 197 Advanced Encryption Standard (AES). November 26, 2001; Updated May 9, 2023

[30] Delgado-Rajo F, Melian-Segura A, Guerra V, Perez-Jimenez R, Sanchez-Rodriguez D. Hybrid RF/VLC network architecture for the internet of things. Sensors. 2020;**20**:478. DOI: 10.3390/s20020478

[31] IEEE Computer Society LAN/MAN Standards Committee, Institute of Electrical and Electronics Engineers, IEEE-SA Standards Board. NJ, USA: IEEE; 2011. DOI: 10.1109/IEEESTD.2011.6016195

[32] Ghassemlooy Z, Luo P, Zvanovec S. Optical camera communications. In: Uysal M, Capsoni C, Ghassemlooy Z, Boucouvalas A, Udvary E, editors. Optical Wireless Communications: Signals and Communication Technology. Cham: Springer; 2016. DOI: 10.1007/978-3-319-30201-0_25

[33] Peruzzi G, Pozzebon A. Combining LoRaWAN and NB-IoT for edge-to-cloud low power connectivity leveraging on fog computing. Applied Sciences. 2022;**12**:1497. DOI: 10.3390/app12031497

[34] MQTT Protocol. Available from: http://mqtt.org [Accessed: 13 December 2021]

[35] Mikhaylov K, et al. Multi-RAT LPWAN in Smart Cities: Trial of LoRaWAN and NB-IoT Integration. 2018 IEEE International Conference on Communications (ICC), Kansas City, MO, USA; 2018. pp. 1-6. DOI: 10.1109/ICC.2018.8422979

[36] Puspitorini O, Wijayanti A, Mahmudah H, Siswandari N. Analysis of Lightweight Boat Communication Equipment Performance for Data Transmission onVHF Long Range (LoRa) Onshore Network System. In: 2022 International electronics symposium (IES). Surabaya, Indonesia: IES; 2022. pp. 227-231. DOI: 10.1109/IES55876.2022.9888670

[37] Jovalekic N, Drndarevic V, Pietrosemoli E, Zennaro IDAM. Experimental study of LoRa transmission over seawater. Sensors (Basel). 2018;**18**(9):2853. DOI: 10.3390/s18092853

Chapter 8

IoT and Energy

Mohammed M. Alenazi

Abstract

The Internet of Things (IoT) has the potential to revolutionize energy management by enabling the collection and analysis of real-time data from various energy sources. This research paper investigates the impact of the Internet of Things (IoT) on energy management. The paper provides an overview of IoT and its potential applications in energy management, including improved efficiency, reduced costs, and better resource utilization. The benefits of using IoT for energy management and the major challenges that may arise in implementing IoT-enabled energy management are discussed. Potential solutions to these challenges, such as artificial intelligence and cloud computing, are presented, along with case studies of IoT-enabled energy management in different industries. The paper also analyzes the impact of IoT on energy efficiency in telecommunications and cloud infrastructure. Finally, the future outlook for IoT and energy management is discussed, including potential developments in edge computing, advanced analytics, and 5G networks. Overall, this paper highlights the potential of IoT to revolutionize energy management and provides insights into the challenges and opportunities of implementing IoT-enabled energy management solutions.

Keywords: Internet of Things, IoT, energy management, energy efficiency, artificial intelligence, cloud computing, edge computing, advanced analytics, 5G networks, case studies

1. Introduction

The Internet of Things (IoT) is a rapidly growing technology transforming many industries, including energy management. IoT has the potential to revolutionize the way we manage energy by enabling real-time monitoring, analysis, and control of energy consumption. This research paper explores IoT's impact on energy management and the potential benefits and challenges of implementing IoT-enabled energy management solutions.

IoT refers to the interconnectedness of devices, sensors, and other objects that can communicate and exchange data over the internet [1]. IoT-enabled devices can collect and transmit vast amounts of data, which can be analyzed using artificial intelligence and other advanced analytics tools to optimize energy consumption. The increase in the numbers of the Internet of Things in the future and its connection to the Internet increases the amount of data that created and needs to process, leading to an increase in the energy in the devices to process those data [2]. It is predicted that the quantity of Internet of Things (IoT) gadgets across the globe will increase nearly threefold, rising from 9.7 billion in 2020 to over 29 billion in 2030 [2].

Energy management is the process of monitoring, controlling, and conserving energy usage in buildings, factories, transportation, and other sectors. IoT in energy management can provide many benefits, such as improved efficiency, reduced costs, and better resource utilization. IoT-enabled energy management solutions can also help reduce greenhouse gas emissions and contribute to a more sustainable future.

This research paper aims to provide a comprehensive overview of the impact of IoT on energy management. The objectives of the paper are to:

- Describe the benefits of IoT in energy management and provide examples of how IoT can be used to optimize energy consumption and reduce waste.

- Identify the significant challenges of using IoT for energy management and discuss potential solutions to address these challenges.

- Present case studies of IoT-enabled energy management in different industries.

- Analyze the impact of IoT on energy efficiency in telecommunications and cloud infrastructure.

- Discuss the potential future developments in IoT and energy management, including edge computing, advanced analytics, and 5G networks.

2. Overview of the Internet of Things (IoT)

The Internet of Things (IoT) is a network of connected devices, objects, and sensors that can collect and exchange data without human intervention. IoT devices are typically embedded with sensors, software, and other technologies that enable them to communicate with each other and other devices, such as smartphones or computers.

IoT has several key characteristics that distinguish it from traditional computing systems. Firstly, IoT devices are highly interconnected, allowing them to share data and collaborate. Secondly, IoT devices are typically embedded in everyday objects, such as appliances, vehicles, and buildings, making them highly pervasive. Thirdly, IoT devices are often low-power and low-cost, making them accessible to many users.

Figure 1 provides a graphical representation of the evolution in the number of connected devices within the Internet of Things (IoT) ecosystem over the years, spanning from 2003 to a projection for 2025. The figure visually conveys the exponential growth in connected devices, underscoring the rapid expansion of the IoT landscape. The data captures the remarkable rise in device connectivity, indicating the trend's trajectory over time. By illustrating this growth pattern, **Figure 1** emphasizes the IoT's significant role in transforming how devices interact and share data, further driving the potential for enhanced energy management and efficiency.

IoT uses sensors to collect data from the environment, and then transmit it over a wireless network to a cloud-based platform for analysis and storage. IoT devices can be controlled and monitored remotely through a smartphone or computer, allowing users to adjust settings or receive real-time notifications.

IoT has various applications in various industries, such as manufacturing, healthcare, transportation, and energy management. IoT can optimize production

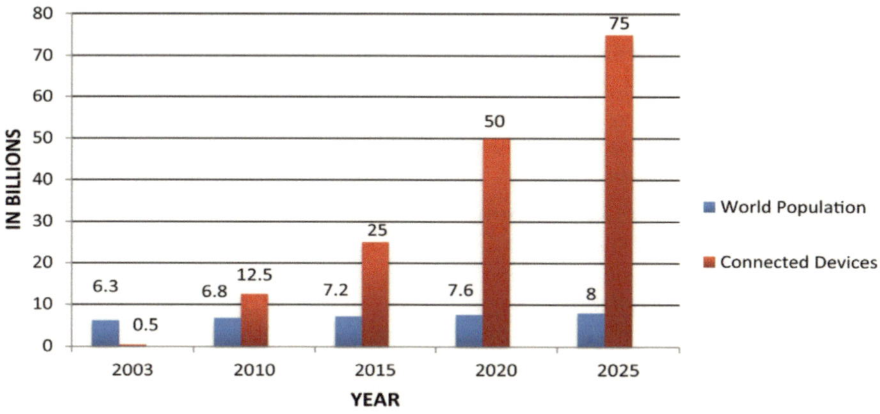

Figure 1.
Number of connected devices on the Internet of Things (2003–2025) [3].

processes, monitor equipment performance, and reduce downtime in manufacturing. IoT can monitor patients remotely, track medication adherence, and improve patient outcomes in healthcare. IoT can improve logistics, reduce traffic congestion, and enhance driver safety in transportation [4]. In energy management, IoT can monitor and control energy usage in buildings, factories, and other settings, optimizing energy consumption and reducing waste.

Overall, IoT has the potential to transform many industries by enabling real-time monitoring, analysis, and control of data. The following sections of this research paper will explore how IoT can optimize energy management and the challenges and opportunities of implementing IoT-enabled energy management solutions.

Figure 2 visually presents the distribution of the Internet of Things (IoT) market across different subsectors in the year 2017. The figure showcases the varying market shares of IoT in distinct industries, offering insights into the sectors that were adopting IoT solutions at that time. This data aids in understanding the prevalence of IoT across sectors such as manufacturing, healthcare, transportation, and energy management. By visually representing the distribution of IoT market

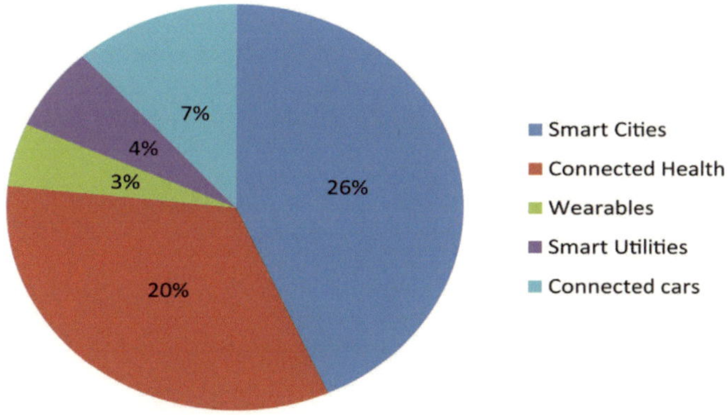

Figure 2.
Internet of Things global market share by subsector (2017) [3].

share, **Figure 2** highlights the diverse applications and potential impact of IoT technology across various industries.

3. Benefits of IoT for energy management

IoT in energy management offers many benefits, including improved efficiency, reduced costs, and better resource utilization. IoT-enabled devices can collect vast amounts of data on energy usage, which can be analyzed in real-time using advanced analytics tools to optimize energy consumption and reduce waste. The following are some specific examples of how IoT can be used to improve energy management (**Figure 3**).

3.1 Real-time monitoring and control

IoT devices can provide real-time data on energy usage, enabling users to monitor and control energy consumption [6]. This can help identify areas of energy waste and optimize energy usage, reducing costs and improving efficiency.

3.2 Predictive maintenance

IoT devices can also monitor equipment performance and identify potential issues before they occur [7]. This can help prevent downtime, reduce maintenance costs, and optimize energy usage.

3.3 Demand response

IoT can implement demand response programs, incentivizing consumers to reduce energy usage during peak demand periods. This can help reduce strain on the energy grid and prevent blackouts while reducing consumer costs.

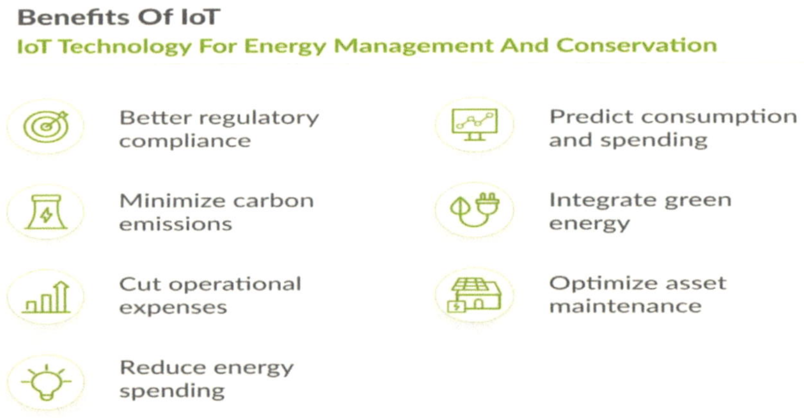

Figure 3.
Benefits of IoT for energy management [5].

3.4 Energy storage

IoT-enabled energy storage systems can store excess energy generated by renewable sources, such as solar or wind power, for later use [8]. This can help reduce reliance on fossil fuels and promote renewable energy sources.

3.5 Smart grid optimization

IoT can optimize the energy grid by monitoring and controlling energy distribution in real-time [6]. This can help reduce energy waste, improve efficiency, and prevent blackouts.

IoT in energy management can provide many benefits, helping reduce costs, improve efficiency, and promote sustainability. The next section of this research paper will explore the challenges of implementing IoT-enabled energy management solutions and potential solutions to address these challenges.

The benefits of integrating the Internet of Things (IoT) into energy management are underscored by compelling real-world examples. For instance, McKinsey's research reveals that IoT-enabled energy management systems in commercial buildings could yield energy consumption reductions of 15–20% and operational cost savings of 10–15%. General Electric's implementation of an IoT-based energy management system in a manufacturing facility resulted in an impressive 10% reduction in energy consumption within the first year of deployment. Moreover, Vodafone's adoption of an IoT-powered smart meter solution led to a notable 12% reduction in energy consumption across its commercial properties. Such examples vividly illustrate the potential for IoT to drive substantial efficiency gains in energy management.

Despite these benefits, challenges associated with IoT implementation in energy management should not be underestimated. Deloitte's survey findings indicate that 48% of respondents identified data security as a substantial challenge in implementing IoT-enabled energy management solutions. The World Economic Forum's perspective on interoperability issues is equally noteworthy, suggesting that discrepancies between IoT devices and existing energy infrastructure might lead to up to $120 billion in lost value by 2025. Addressing these concerns is essential, given the significant stakes involved. A report by the Industrial Internet Consortium raises alarms about the absence of standardized security protocols for IoT devices, potentially exposing critical energy infrastructure to cyber threats. With the International Data Corporation estimating an impending surge in IoT device connectivity—possibly reaching 45 billion devices by 2023—the potential attack surface for cyber-attacks is set to expand significantly.

Real-world case studies further illustrate IoT's prowess in energy management. Johnson Controls' implementation of an IoT-based energy management system in a hospital stands out, with a remarkable 22% reduction in energy consumption and annual cost savings totaling $2.2 million. A city renowned for its smart city initiatives, Barcelona, successfully deployed IoT-enabled smart street lighting, resulting in a commendable 30% decrease in energy consumption and an equally noteworthy 35% reduction in maintenance costs. The application of IoT-driven solutions extends beyond urban settings: Siemens' development of an IoT-based solution for wind farm optimization enhanced the efficiency of wind turbines, leading to a noteworthy 10–20% increase in energy output. These cases spotlight the transformative potential of IoT in diverse energy management contexts.

The impact of IoT on energy efficiency is not confined to specific sectors but extends to telecommunications and cloud computing. Ericsson's research accentuates this by suggesting that IoT-enabled energy management solutions within telecommunications networks could translate into substantial energy savings of up to 40%. Furthermore, Google's successful integration of artificial intelligence (AI) and IoT for data center energy management resulted in an impressive 15% reduction in overall energy consumption. A promising projection by Cisco underscores the positive trajectory of IoT's influence: it estimates that IoT devices connected to 5G networks may yield energy consumption reductions of up to 90% compared to traditional cellular networks. These examples highlight the cross-industry potential for IoT to foster energy efficiency and sustainability.

4. Challenges of IoT for energy management

Despite the potential benefits of using IoT in energy management, several challenges must be addressed. The following are some of the significant challenges of using IoT for energy management (**Figure 4**).

4.1 Security risks

IoT devices can be vulnerable to cyber-attacks, posing significant security risks. These risks can include data breaches, theft of intellectual property, and disruptions to critical infrastructure [10]. Therefore, ensuring the security of IoT-enabled energy management systems is essential to minimize these risks. This can be achieved through secure communication protocols, strong authentication, access control mechanisms, and regular security updates and patches.

4.2 Interoperability issues

IoT devices and systems can be complex and varied, making interoperability between devices and systems difficult. This can create challenges in integrating IoT

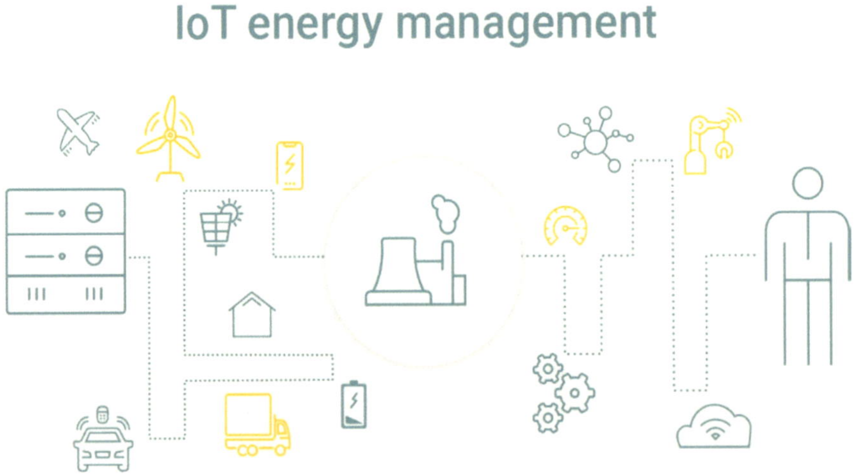

Figure 4.
IoT for energy management [9].

devices with energy management systems, leading to additional costs and complexity [3]. Addressing this challenge requires the development of standard communication protocols and data formats to ensure interoperability between different devices and systems.

4.3 Privacy concerns

IoT devices can collect and transmit large amounts of data, including personal data, raising privacy concerns [10]. Therefore, ensuring the privacy of individuals' data is critical to ensure trust and confidence in IoT-enabled energy management solutions. This can be achieved through data encryption, anonymization techniques, and data minimization strategies.

4.4 Lack of standardization

IoT is still a relatively new technology, and there needs to be more standardization in many areas, including data formats, communication protocols, and security standards. This lack of standardization can challenge IoT-enabled energy management solutions' interoperability, security, and reliability [3]. Addressing this challenge requires the development of common standards and guidelines to ensure consistency and interoperability between different IoT devices and systems.

These challenges can impact the implementation of IoT in energy management by increasing costs, reducing reliability, and decreasing user trust and confidence.

5. Potential solutions for IoT-enabled energy management

Potential solutions have been proposed to address the challenges of IoT-enabled energy management. These include:

5.1 Artificial intelligence (AI)

AI technologies, such as machine learning and predictive analytics, can be used to analyze and interpret the large amounts of data generated by IoT devices, providing insights into energy usage patterns, and identifying areas for optimization. AI can also be used to automate energy management processes, reducing the need for manual intervention, and improving overall efficiency [11]. For example, AI algorithms can predict energy demand and adjust supply, accordingly, ensuring energy resources are used more effectively. Artificial intelligence (AI): AI technologies hold promise in addressing the complexities of IoT-enabled energy management. Machine learning algorithms can analyze the massive volume of data generated by IoT devices, uncovering patterns and trends that might be difficult for humans to identify. Predictive analytics can forecast energy demand based on historical data, weather forecasts, and other factors. AI-driven optimization can dynamically adjust energy consumption patterns, making systems more responsive and efficient. For instance, AI can optimize the operation of smart grids by balancing energy supply and demand, reducing wastage, and minimizing costs. By automating energy management processes, AI reduces human error and allows for real-time decision-making, resulting in more effective resource allocation.

5.2 Blockchain

Blockchain technology can improve the security and privacy of IoT-enabled energy management systems. Using a decentralized, tamper-proof ledger, blockchain can help prevent unauthorized access and data manipulation, ensuring the integrity and security of energy management systems [12]. Additionally, blockchain can facilitate secure, peer-to-peer energy transactions, enabling more efficient and flexible energy management solutions (**Figure 5**).

Blockchain technology addresses critical concerns related to data security, privacy, and transparency in IoT-enabled energy management. By creating an immutable and transparent record of energy transactions and data exchanges, blockchain ensures data integrity and prevents unauthorized tampering. Decentralized and peer-to-peer energy trading becomes feasible, enabling direct transactions between energy producers and consumers. This disintermediation can lead to more efficient utilization of energy resources and greater flexibility in energy management. Blockchain's auditability also aids regulatory compliance, which is crucial in regulated energy markets. Energy certificates, carbon credits, and other compliance-related records can be securely stored and verified on the blockchain.

5.3 Cloud computing

Cloud computing can enhance the scalability and reliability of IoT-enabled energy management systems. By providing on-demand access to computing resources and storage, cloud computing can help to process and analyze the large amounts of data generated by IoT devices, enabling more effective energy management [14]. Additionally, cloud computing can help improve energy management systems' reliability and resilience by providing redundant storage and computing resources (**Figure 6**).

Cloud computing offers scalable computational resources to handle the data-intensive demands of IoT-enabled energy management systems. The cloud provides the necessary infrastructure for processing and analyzing the massive amounts of

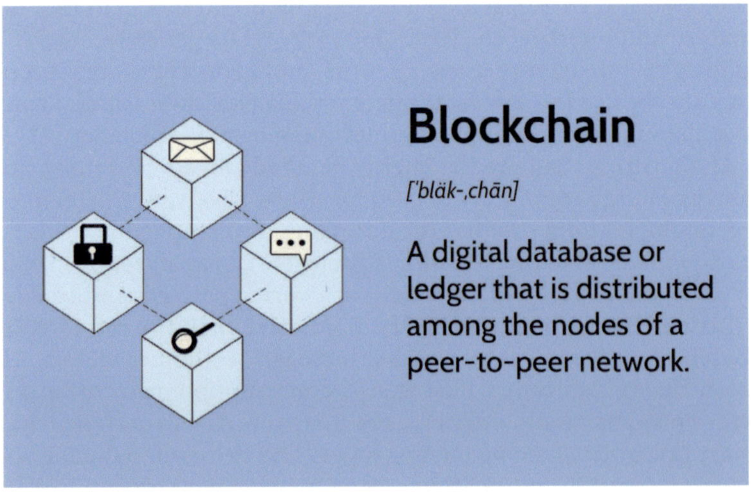

Figure 5.
Blockchain [13].

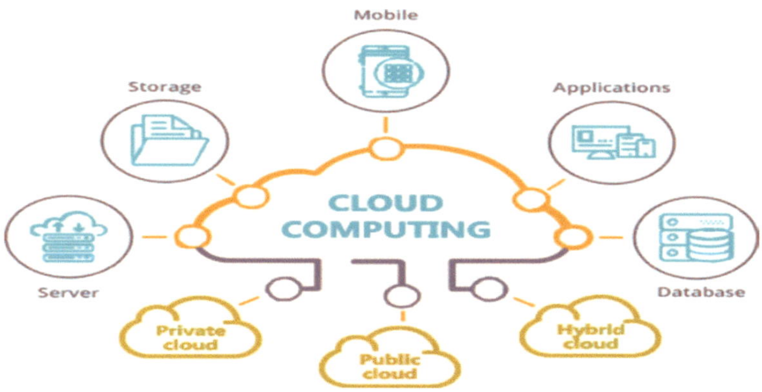

Figure 6.
Cloud computing [15].

data generated by IoT devices in real-time. This enables continuous monitoring, rapid response to changing conditions, and data-driven insights for better decision-making. Moreover, cloud-based solutions enhance system reliability and resilience by offering redundant storage and computational resources. The ability to remotely access and control energy management systems through the cloud empowers users to monitor and adjust energy consumption even from remote locations, contributing to overall efficiency.

Figure 7 shows the collaboration between cloud computing and the Internet of Things (IoT) for energy management optimization. Cloud computing is depicted as a central hub, offering scalable resources and storage. This supports real-time processing of data from IoT devices, ensuring continuous monitoring and informed decision-making. IoT devices gather data via sensors, transmitting it wirelessly to the cloud for analysis. This synergy enables efficient energy management, thanks to the cloud's computational capacity.

Examples of how these solutions can be used to improve energy management efficiency and effectiveness include:

Using AI algorithms to optimize energy consumption in buildings by analyzing data on occupancy patterns, weather conditions, and energy usage to adjust heating, lighting, and ventilation systems.

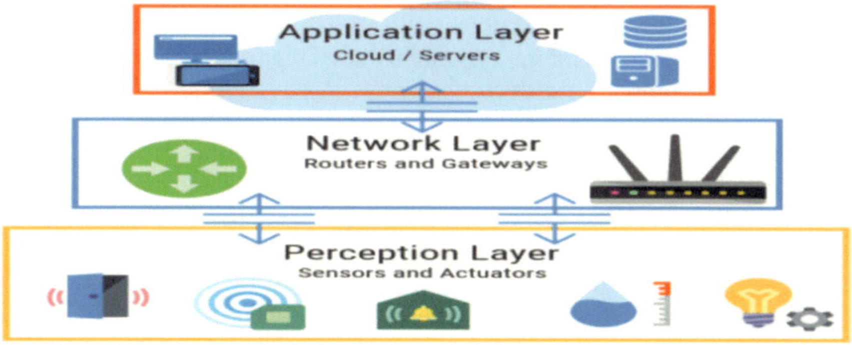

Figure 7.
Cloud-IoT infrastructure [16].

Using blockchain technology to enable secure, peer-to-peer energy trading between households and businesses, enabling more efficient and flexible energy management solutions.

Using cloud computing to process and analyze data from IoT-enabled energy management systems enables real-time monitoring and control of energy usage and improves overall system reliability and efficiency.

By leveraging these potential solutions, IoT-enabled energy management systems can be made more secure, reliable, and efficient, overcoming the challenges identified in the previous section and enabling the realization of the full potential of IoT in energy management.

Examples of implementing solutions:

Several practical examples highlight the potential benefits of these solutions:

- AI for building energy optimization: AI algorithms can optimize energy consumption in buildings by analyzing occupancy patterns, weather data, and historical energy usage. Heating, cooling, and lighting systems can be automatically adjusted based on real-time conditions, leading to energy savings without compromising comfort.

- Blockchain-powered peer-to-peer energy trading: Blockchain enables households and businesses to directly trade excess energy with each other. This decentralized approach eliminates intermediaries, reducing transaction costs and promoting energy efficiency through more localized energy distribution.

- Cloud-enhanced real-time monitoring: Cloud computing allows for real-time monitoring of energy consumption across multiple locations. This enables facility managers to identify inefficiencies promptly and take corrective actions, resulting in reduced energy waste.

5.4 Expanding on limitations and challenges

Continuing the discussion on the challenges of implementing AI, blockchain, and cloud computing in energy management:

- AI: While AI offers predictive and optimization capabilities, its effectiveness heavily relies on data quality. Inaccurate or incomplete data can lead to suboptimal recommendations. Developing and training AI models also demand significant computational resources, potentially offsetting energy savings. Moreover, ensuring transparency and interpretability of AI algorithms is crucial to gaining user trust and complying with regulations.

- Blockchain: Despite its security benefits, blockchain's computational demands and scalability issues pose challenges. Traditional blockchain consensus mechanisms like Proof of Work can consume substantial energy. Achieving consensus in real-time energy transactions might not align with the energy efficiency goals of such systems. Additionally, regulatory alignment and privacy concerns need careful attention when integrating blockchain into existing energy frameworks.

- Cloud computing: While cloud solutions offer scalability and remote access, they introduce latency due to internet connectivity. This latency can impact real-time decision-making, critical for efficient energy management. Data security remains a concern, as sensitive energy consumption data stored in the cloud could be vulnerable to breaches. Moreover, the financial implications of long-term cloud service usage, coupled with potential vendor lock-in, require consideration.

Common challenges: Overcoming resistance to change and fostering interdisciplinary collaboration is essential. Organizations must navigate complex regulatory landscapes and ensure compliance with data protection laws. Comprehensive cost–benefit analyses are vital to assess the viability of these technologies in the context of energy savings. Ultimately, addressing these challenges demands a holistic approach that balances technological innovation, energy efficiency goals, and practical considerations.

While AI, blockchain, and cloud computing offer solutions to IoT-enabled energy management challenges, their successful integration require careful consideration of their respective limitations and the unique demands of energy systems. By thoughtfully applying these technologies, energy management can be transformed, enhancing efficiency, security, and sustainability.

6. Case studies of IoT-enabled energy management

Several case studies have examined the implementation of IoT-enabled energy management in various industries. The following are examples of such case studies:

6.1 Smart homes

A systematic study investigated the impact of smart home technology on energy consumption. The study utilized IoT-enabled devices such as smart thermostats, smart lighting, and smart appliances. The devices were connected to a central hub, providing real-time energy consumption data. The results showed a significant reduction in energy consumption by 10–15%, resulting in cost savings for homeowners [17]. However, the study also highlighted the need for data security and privacy measures to be implemented, as the devices collected sensitive information such as occupancy patterns. The study by Alenazi et al. proposes an energy-efficient neural network embedding technique in IoT over passive optical networks to enhance the performance of IoT-based applications while reducing energy consumption [18].

6.2 Cloud distribution

In addition to using AI and IoT to reduce energy consumption, the work by Alenazi et al. also proposes the concept of cloud distribution to enhance energy efficiency in IoT-based applications further. The cloud distribution approach involves distributing the processing load of an IoT application across multiple virtual machines in the cloud. This allows the load to be balanced across multiple machines, reducing the energy consumption of each machine and increasing overall efficiency [19]. The study evaluates the energy efficiency of this approach in comparison to traditional approaches and demonstrates its effectiveness in reducing energy consumption [19]. With the increasing demand for IoT-based services and the associated energy consumption, the proposed approach can significantly benefit both the environment and the economic sustainability of IoT-based applications.

6.3 Manufacturing

In another study, IoT-enabled sensors were installed in a manufacturing plant to monitor energy consumption. The sensors were placed on equipment such as motors, compressors, and conveyors and provided real-time data on energy usage [20]. The data were analyzed using machine learning algorithms, which identified areas for optimization. Implementing IoT-enabled energy management significantly reduced energy consumption annually. However, the study also highlighted the need for interoperability among different IoT systems, as the sensors used in the study were from different manufacturers, and it was challenging to integrate the data from these sensors.

6.4 Transportation

The California Department of Transportation (Caltrans) implemented an IoT-enabled energy management system in its highway lighting systems [21]. The system utilized sensors and smart lighting technology to monitor and control lighting usage. The sensors on the light poles provided real-time energy consumption data. The system was also integrated with weather data, which enabled the lighting to be adjusted based on ambient light levels. Implementing the system resulted in a reduction in energy consumption and a significant reduction in maintenance costs [21]. However, the study also highlighted the need for data security measures, as the system collected sensitive information such as traffic patterns and vehicle speeds.

In these case studies, IoT-enabled energy management systems provided significant benefits, such as improved efficiency, cost savings, and better resource utilization. However, implementing these systems posed several challenges, including data security risks, interoperability issues, and privacy concerns. For example, in the case of the smart home study, concerns were raised regarding the security and privacy of the data collected by the IoT-enabled devices. Therefore, it is crucial to address these challenges to realize the potential benefits of IoT-enabled energy management systems.

Figure 8 illustrates a visual representation of diverse applications of the Internet of Things (IoT). These applications encompass various sectors, showcasing IoT's wide-ranging impact. The figure depicts IoT's transformative influence on sectors such as smart homes, agriculture, healthcare, manufacturing, transportation, and energy management. In smart homes, IoT-enabled devices, including thermostats,

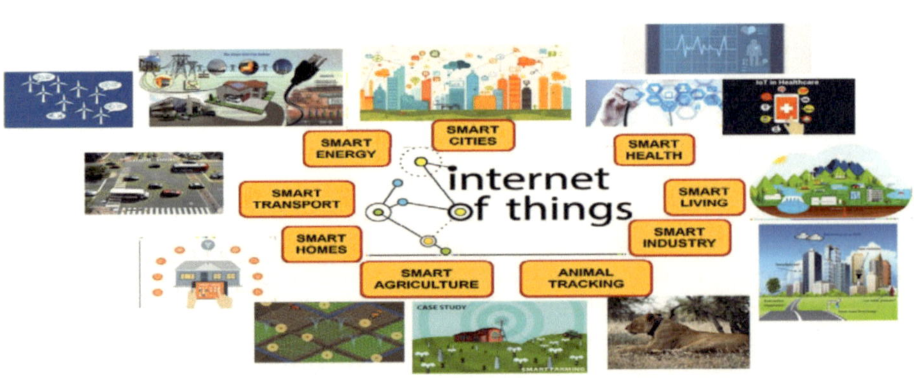

Figure 8.
IoT applications [4].

lighting, and appliances, connect and exchange data to optimize energy consumption and enhance convenience. In agriculture, sensors integrated into fields and livestock enable real-time monitoring of conditions, facilitating precision farming and resource optimization. In healthcare, IoT devices collect patient data, enable remote monitoring, and enhance medical diagnostics and treatment. In manufacturing, IoT-driven monitoring and automation optimize production processes and equipment maintenance. In transportation, connected vehicles and smart traffic systems improve traffic flow, reduce congestion, and enhance driver safety. Importantly, the figure highlights the specific application of IoT in energy management, demonstrating its potential to revolutionize the efficient use of energy resources across sectors.

IoT's transformative potential extends beyond energy management into telecommunications and cloud computing. Ericsson's research suggests potential energy savings of up to 40% through IoT-enabled energy management in telecommunications networks. Moreover, Google's integration of AI and IoT in data center energy management achieved a commendable 15% reduction in energy consumption. Cisco's projection of IoT devices connected to 5G networks yielding up to 90% energy consumption reduction underscores IoT's cross-industry energy efficiency potential.

7. Regulatory challenges and policy frameworks

As IoT revolutionizes energy management, regulatory challenges emerge alongside policy frameworks' necessity to support its growth. The intricate interplay between energy consumption, data privacy, and security mandates comprehensive guidelines. Regulatory hurdles arise from IoT-enabled energy management innovation. Balancing data collection, individual privacy rights, and security requirements is crucial. Robust policies must establish standards for data security, encryption, and access control to safeguard critical infrastructure while fostering innovation. Collaborative efforts between government bodies and industry associations are essential to define regulations aligning innovation with security and privacy imperatives.

8. Future outlook for IoT and energy management

Anticipating IoT's trajectory reveals three pivotal areas: edge computing, advanced analytics, and 5G networks. Edge computing's promise lies in faster processing and reduced energy consumption. However, its implementation requires substantial infrastructure investment and standardized protocols. Advanced analytics, driven by AI, holds the potential for identifying energy consumption patterns. Ensuring data accuracy and developing skilled personnel are prerequisites. Integrating 5G networks promises enhanced communication, albeit entailing regulatory and infrastructural challenges.

9. Energy efficiency of IoT in telecommunications and cloud

IoT is transforming the telecommunications and cloud computing industries, increasing energy efficiency and sustainability.

Telecommunications networks have traditionally been designed to handle peak loads, resulting in inefficient energy usage during periods of low usage. However, integrating IoT sensors and analytics can help operators optimize network management, reducing energy consumption and costs [22]. IoT sensors can monitor network activity, detect anomalies, and adjust network capacity in real-time. This helps minimize energy consumption while maintaining the required network performance and reliability levels.

One successful example of IoT in telecommunications is Vodafone's IoT-enabled smart meter solution [23]. The solution includes smart meters that collect real-time energy consumption data, allowing Vodafone to identify energy inefficiencies and develop strategies to optimize energy usage. As a result, Vodafone has reduced energy consumption, leading to significant cost savings.

Data centers are responsible for significant energy consumption in the cloud computing industry. IoT can help reduce energy consumption by optimizing server utilization and cooling systems. IoT sensors can monitor server utilization and adjust capacity, accordingly, powering down unused servers to save energy. Additionally, IoT sensors can monitor cooling systems and adjust them to optimize energy consumption.

One successful example of IoT in cloud computing is Microsoft's IoT-enabled data center cooling system. Microsoft installed sensors to monitor temperature and humidity levels in their data centers, allowing the company to adjust the cooling system to optimize energy consumption. The result was a 30% reduction in energy consumption, leading to significant cost savings and environmental benefits.

However, implementing IoT in telecommunications and cloud computing infrastructure also poses challenges. One major challenge is ensuring the security of IoT devices and networks. As the number of connected devices increases, the attack surface expands, increasing the risk of cyber threats. It is essential to ensure that IoT devices are secure and networks are protected from cyber-attacks.

10. Future outlook for IoT and energy management

The integration of IoT in energy management has already led to significant benefits, but the future of this technology is even more promising. The following are potential developments that may shape the future of IoT and energy management:

10.1 Edge computing

One potential development is the use of edge computing. Edge computing involves processing data closer to the source, resulting in faster processing times and reduced energy usage. By processing data at the edge, it is possible to reduce the amount of data that needs to be transmitted to the cloud, leading to lower energy consumption and reduced latency. However, implementing edge computing may require significant investment in infrastructure and the adoption of standard protocols for interoperability and security.

10.2 Advanced analytics

Advanced analytics, such as machine learning and predictive analytics, can help identify energy usage patterns and optimize energy consumption. With the help of advanced analytics, it is possible to identify trends in energy usage and adjust energy consumption accordingly. This can lead to significant energy savings and reduce waste. However, implementing advanced analytics may require significant investment in data collection, processing, and analysis and require highly skilled personnel.

10.3 Integration of 5G networks

Another potential development is the integration of 5G networks. 5G networks can enable faster and more reliable communication between IoT devices and energy management systems. This can lead to better energy management in smart cities, factories, and other industrial applications. However, implementing 5G networks may require significant investment in infrastructure and may face regulatory hurdles related to the allocation of radio spectrum and privacy concerns.

10.4 Blockchain

Blockchain technology can be used to improve energy management. By leveraging distributed ledger technology, it is possible to create a secure and transparent energy marketplace that enables peer-to-peer energy trading [12]. This can lead to more efficient energy distribution and reduced energy costs. However, implementing blockchain technology may require significant investment in infrastructure and may face regulatory hurdles related to integrating decentralized systems into existing centralized energy systems.

However, the implementation of these developments also comes with potential challenges. Below are some of the potential challenges that could arise from future developments in IoT and energy management:

Increased complexity: As IoT systems become more complex, managing them could become increasingly challenging. This could lead to more significant maintenance costs and difficulties in troubleshooting. Managing the complexity of IoT systems will require developing new tools, standards, and procedures.

Security risks: The more devices are connected to a network, the greater the risk of cybersecurity threats [10]. As IoT systems become more widespread, ensuring their security will become increasingly important. The integration of security measures and the development of new protocols will be essential to mitigate these risks.

Interoperability issues: As more devices are connected to an IoT network, ensuring they can communicate effectively could become challenging. Interoperability issues could result in data loss or other inefficiencies. Developing standardized protocols and ensuring the compatibility of devices will be critical to overcoming these challenges.

Data privacy concerns: As IoT systems generate vast amounts of data, there will be growing concerns about how that data is collected, stored, and used. Ensuring the privacy of individual data will be a significant challenge [10]. Addressing privacy concerns will require the development of new policies, regulations, and technologies.

Resource constraints: Batteries or other limited energy sources often power IoT devices. As IoT devices grow, ensuring they are all powered efficiently and sustainably could become a significant challenge. Addressing resource constraints will require the development of new energy sources, storage systems, and power management technologies.

Regulatory challenges: As IoT devices become more ubiquitous, there may be challenges in regulating their use and ensuring that they adhere to applicable legislation, regulations, and standards [24]. Addressing regulatory challenges will require the development of new policies, regulations, and standards that can accommodate the unique features of IoT systems.

As the IoT continues to evolve, numerous opportunities exist to advance energy management systems through emerging technologies. For example, quantum computing could optimize energy consumption by enabling more precise modeling and simulation of energy systems. Distributed ledger technologies, such as blockchain, could also be employed to create secure and transparent energy trading systems that enable more efficient distribution and use of energy.

To fully realize these opportunities, future research can explore these emerging technologies' potential benefits and challenges and identify how they can be integrated into existing energy management systems. Research can also focus on developing best practices for scalability, reliability, and data management in IoT-enabled energy management systems, particularly in large-scale deployments. This includes developing standardized protocols for interoperability and security that enable the seamless integration of various devices and systems.

Another critical area for future research is the development of new business models for IoT-enabled energy management. This could include exploring innovative pricing structures that incentivize energy conservation and reward energy-efficient behavior. It could also involve developing new energy trading and management approaches that exploit emerging technologies such as blockchain.

Overall, there is a need for continued research in IoT-enabled energy management to realize this technology's potential benefits fully. By tackling the issues and opportunities related to the Internet of Things (IoT) and energy consumption, researchers can contribute to developing an efficient and more sustainable energy future.

11. Conclusion

In conclusion, this research paper examined the integration of IoT in energy management and its potential impact on energy efficiency and sustainability. The paper explored the challenges and benefits of IoT-enabled energy management and presented case studies from different industries, such as smart homes, manufacturing, and transportation. It also analyzed the impact of IoT on energy efficiency in telecommunications and cloud infrastructure, as well as future developments in IoT and energy management, such as edge computing, advanced analytics, 5G networks, and blockchain.

The findings of this research paper suggest that IoT has significant potential to improve energy efficiency, reduce energy consumption, and promote sustainability. However, challenges associated with integrating IoT in energy management include scalability, reliability, security, and privacy concerns.

The implications of this research are significant for energy management and IoT applications. IoT in energy management can help reduce energy consumption, optimize energy usage, and promote sustainability. It can also lead to cost savings and reduce carbon emissions. The findings of this research can inform policymakers, energy managers, and industry professionals about the potential benefits and challenges of IoT-enabled energy management and guide the development of best practices and standards.

To further advance research in this field, future studies can focus on identifying best practices for integrating IoT into energy management systems and addressing the challenges associated with scalability, reliability, and data management. Additionally, research can explore emerging technologies, such as quantum computing and distributed ledger technologies, to improve energy management efficiency and effectiveness. Standardized protocols for interoperability and security are also essential to enable the widespread adoption of IoT-enabled energy management.

Overall, this research paper highlights the potential of IoT-enabled energy management to promote sustainability and reduce energy consumption. The findings of this research can inform policymakers, energy managers, and industry professionals about the benefits and challenges of integrating IoT in energy management and guide the development of best practices and standards for future

Acknowledgements

The author would like to express their sincere gratitude to the University of Tabuk for their support and encouragement.

The author would like to thank the University of Tabuk for providing a stimulating research environment and for their support throughout the research process. We are grateful to the University of Tabuk for their generous assistance in the form of resources that we were able to use to conduct our research.

Conflict of interest

No.

Notes/thanks/other declarations

No.

Author details

Mohammed M. Alenazi
University of Tabuk, Tabuk, Saudi Arabia

*Address all correspondence to: m-alenazi@ut.edu.sa

References

[1] Alenazi MM, Yosuf BA, Mohamed SH, El-Gorashi TEH, Elmirghani JMH. Energy-efficient distributed machine learning in cloud fog networks. In: 2021 IEEE 7th World Forum on Internet of Things (WF-IoT). New Orleans, LA, USA; 2021. pp. 935-941. DOI: 10.1109/WF-IoT51360.2021.9595351

[2] Vailshery LS. Number of Internet of Things (IoT) Connected Devices Worldwide from 2019 to 2023, with Forecasts from 2022 to 2030. Statista. Available from: https://www.statista.com/statistics/1183457/iot-connected-devices-worldwide/ [Accessed: September 28, 2023]

[3] Rana B, Singh Y, Singh PK. A systematic survey on internet of things: Energy efficiency and interoperability perspective. Transactions on Emerging Telecommunications Technologies. 2021;**32**(8):e4166

[4] Patel KK, Patel SM, Scholar PG. Internet of things-IOT: Definition, characteristics, architecture, enabling technologies, application and future challenges. International Journal of Engineering Science and Computing. 2016;**6**:5

[5] Galooli. 5 Tips to Save Big on IOT Energy Management. Galooli. Available from: https://galooli.com/blog/iot-energy-management/ [Accessed: September 28, 2023]

[6] Priyadharshini SG, Subramani C, Roselyn JP. An IOT based smart metering development for energy management system. International Journal of Electrical Computing Engineering (IJECE). 2019;**9**(4):3041

[7] Coandă P, Avram M, Constantin V. A state of the art of predictive maintenance techniques. In: IOP Conference Series: Materials Science and Engineering. Vol. 997. No. 1. IOP Publishing; 2020

[8] Pu X, Wang ZL. Self-charging power system for distributed energy: Beyond the energy storage unit. Chemical Science. 2020;**12**(1):34-49

[9] Kwiatkowski M. IOT Energy Management for Low-Energy IOT Devices. Available from: https://www.avsystem.com/blog/iot-energy-management-for-low-energy-iot-devices/ [Accessed: September 28, 2023]

[10] Tobby S. Chapter seven: Critical infrastructure and the internet of things. Cyber Security in a Volatile World. 2017. p. 93

[11] Rochd A et al. Design and implementation of an AI-based & IoT-enabled home energy management system: A case study in Benguerir — Morocco. Energy Reports. 2021;**7**:699-719

[12] Kumar NM, Mallick PK. Blockchain technology for security issues and challenges in IoT. Procedia Computer Science. 2018;**132**:1815-1823

[13] Hayes A. Blockchain Facts: What is it, How it Works, and How it Can be Used. Investopedia. Available from: https://www.investopedia.com/terms/b/blockchain.asp [Accessed: September 28, 2023]

[14] Yosuf BA, Mohamed SH, Alenazi MM, El-Gorashi TEH, Elmirghani JMH. Energy-efficient AI over a virtualized cloud fog network. In: Proceedings of the 12th ACM International Conference on Future Energy Systems (e-Energy '21). New York, NY, USA: Association for

Computing Machinery; 2021. pp. 328-334. DOI: 10.1145/3447555.3465378

[15] Narayan A. Know about cloud computing architecture. Knoldus Blogs. Available from: https://blog.knoldus.com/know-about-cloud-computing-architecture/ [Accessed: September 28, 2023]

[16] Sadeeq MM, Abdulkareem NM, Zeebaree SR, Ahmed DM, Sami AS, Zebari RR. IoT and cloud computing issues, challenges and opportunities: A review. Qubahan Academic Journal. 2021;**1**(2):1-7

[17] Issi HN. A Systematic Study of Data Security Issues in Smart Home IoT Devices. [Dissertation] The University of Alabama at Birmingham; 2022

[18] Alenazi MM, Yosuf BA, El-Gorashi T, Elmirghani JMH. Energy efficient neural network embedding in IoT over passive optical networks. In: 2020 22nd International Conference on Transparent Optical Networks (ICTON). Bari, Italy; 2020. pp. 1-6. DOI: 10.1109/ICTON51198.2020.9203403

[19] Alenazi MM, Yosuf BA, Mohamed SH, El-Gorashi TEH, Elmirghani JMH. Energy efficient placement of ML-based services in IoT networks. In: 2022 IEEE International Mediterranean Conference on Communications and Networking (MeditCom). Athens, Greece; 2022. pp. 19-24. DOI: 10.1109/MeditCom55741.2022.9928668

[20] Ma S, Zhang Y, Liu Y, Yang H, Lv J, Ren S. Data-driven sustainable intelligent manufacturing based on demand response for energy-intensive industries. Journal of Cleaner Production. 2020;**274**:123155

[21] Zantalis F, Koulouras G, Karabetsos S, Kandris D. A review of machine learning and IoT in smart transportation. Future Internet. 2019;**11**(4):94

[22] Kamruzzaman MM. Key technologies, applications and trends of internet of things for energy-efficient 6G wireless communication in smart cities. Energies. 2022;**15**(15):5608

[23] Auton N. Internet of things as a business strategy. Scope (Flexible Learning). 2019;**4**:98-103

[24] Michie S, Yardley L, West R, Patrick K, Greaves F. Developing and evaluating digital interventions to promote behavior change in health and health care: Recommendations resulting from an international workshop. Journal of Medical Internet Research. 2017;**19**(6):e232

Chapter 9

Effective Screening and Face Mask Detection for COVID Spread Mitigation Using Deep Learning and Edge Devices

Xishuang Dong, Lucy Nwosu, Sheikh Rufsan Reza and Xiangfang Li

Abstract

The emergence of COVID-19, stemming from the SARS-CoV-2 virus, has led to a widespread outbreak affecting countless individuals and inducing dire circumstances globally. Mitigating the transmission of COVID-19 has necessitated the implementation of effective measures such as rigorous COVID screening and physical safeguards, including practices like social distancing and the utilization of face masks. Notably, the application of advanced technologies such as deep learning, a subset of artificial intelligence (AI), has played a pivotal role in devising novel strategies for both detecting COVID-19 and curbing its propagation. This chapter presents a comprehensive overview of COVID screening methodologies based on deep learning, with a specific focus on biomedical image processing and the detection of face masks. Furthermore, it delves into initial endeavors concerning COVID image analysis and the creation of a mobile face mask detection system, designed to operate on edge devices. The ensuing discussions encompass detailed case studies, showcasing the practical implications and efficacy of these initiatives.

Keywords: COVID-19 pandemic, chest X-ray imaging, deep learning, face mask detection, Internet of Things

1. Introduction

The global COVID-19 pandemic, originating from the Coronavirus disease 2019 (COVID-19), continues to persist across more than 200 countries and territories, evoking significant apprehension within the international community [1, 2]. This crisis has engendered profound human losses and economic ramifications, reshaping lives in various countries through the implementation of lockdown measures. In order to curb the dissemination of COVID-19 and curtail its associated mortality, early detection of the disease assumes critical importance. Effective and prompt screening and testing are pivotal for the proficient management of individuals afflicted by

COVID-19 [3, 4]. The evolution of advanced techniques has led to the proposition of increasingly efficient screening technologies, aimed at attenuating the transmission of COVID-19.

1.1 COVID-19 screening

The goal of a COVID-19 screening test is to identify potential cases of the disease in individuals who are asymptomatic. This proactive approach aims to mitigate the spread of COVID-19 by detecting infections early, allowing for timely and effective treatment. During the initial stages of the COVID-19 pandemic, the primary method for detecting viral infections was the Reverse Transcription-Polymerase Chain Reaction (RT-PCR) test [5]. However, the effectiveness of the RT-PCR assay has been questioned due to its limited sensitivity [6], which can be attributed to various factors like sample preparation and quality control issues [7]. Furthermore, the current nucleic acid tests' sensitivity necessitates repeated testing for a significant portion of suspected patients to achieve a reliable diagnosis. This underscores the need for the development of a complementary tool capable of providing lung-imaging information. Such a tool would serve as an invaluable resource for medical professionals, aiding them in enhancing the accuracy of COVID-19 diagnoses.

Chest X-rays and thoracic computed tomography (CT) scans are readily available imaging tools that offer significant support to medical practitioners in diagnosing lung-related ailments [8–10]. The application of artificial intelligence (AI) in enhancing image analysis of chest X-rays and thoracic CT data has garnered substantial interest, particularly in the development of effective COVID-19 screening techniques. AI, a burgeoning technology in the realm of medical imaging, has played a dynamic role in the battle against COVID-19 [11]. This is in contrast to traditional imaging processes that heavily rely on human interpretation, as AI offers imaging solutions that are safer, more accurate, and more efficient. Notably, the utilization of deep learning for data representation has exhibited remarkable success in image processing [12]. Convolutional neural networks (CNNs) [13–18] have effectively tackled the challenge of representing digital images, particularly on extensive datasets such as the ImageNet dataset [19]. These advances demonstrate the potential of deep learning in transforming biomedical image analysis.

The potency of deep learning methods, such as Convolutional Neural Networks (CNNs), has been prominently demonstrated in the classification of COVID-19 cases. Ghoshal et al. [20] introduce a Bayesian Convolutional Neural Network designed to estimate diagnostic uncertainty in COVID-19 predictions. This approach incorporates 70 lung X-ray images from COVID-19 patients sourced from an online COVID-19 dataset [21], as well as non-COVID-19 images from Kaggle's Chest X-Ray Images (Pneumonia), where Bayesian inference is employed to enhance detection accuracy. Narin et al. [10] focus on COVID-19 infection detection using X-ray images, employing a comparative analysis of three deep learning models: ResNet50, InceptionV3, and Inception-ResNetV2. The evaluation results indicate that the ResNet50 model surpasses the performance of the other two models. Zhang et al. [22] similarly harness the ResNet architecture for COVID-19 classification using X-ray images. An anomaly score is estimated to optimize the COVID-19 score, which in turn is used for classification. Wang et al. [23] introduce COVID-Net, a framework tailored for the detection of COVID-19 cases through X-ray images. Primarily, most ongoing studies employ X-ray images for discriminating between COVID-19 cases and other instances of pneumonia and healthy subjects. However, the limited quantity of available COVID-19 images raises concerns regarding the methods' robustness and generalizability, urging further investigation.

Furthermore, it is of utmost importance to delineate the areas affected by COVID-19 infection, as this yields comprehensive insights crucial for accurate diagnosis. Semantic segmentation plays a pivotal role in identifying and quantifying COVID-19 by recognizing regions and associated patterns. This technique enables the assessment of regions of interest (ROIs) encompassing lung structures, lobes, bronchopulmonary segments, and infected regions or lesions within chest X-ray or CT images. Extracting handcrafted or learned features for diagnosis and other applications becomes feasible through the use of segmented regions. The advancement of deep learning has significantly propelled the evolution of semantic image segmentation. In the context of CT scans, the networks employed for COVID-19 include established models like U-Net [24–26], UNet++ [27], and VB-Net [28] to segment ROIs. Furthermore, segmentation approaches for COVID-19 can be categorized into two main groups: those oriented toward lung regions and those aimed at lung lesions. The former group focuses on distinguishing lung regions, encompassing entire lungs and lung lobes, from surrounding (background) regions in CT or X-ray images [29, 30]. For instance, Jin et al. [29] utilize UNet++ to detect the entire lung region. The latter group seeks to isolate lung lesions (or artifacts such as metal and motion) from lung regions [31, 32]. In addition to screening techniques, physical solutions for mitigating the spread of COVID-19 also hold efficacy.

1.2 Face mask detection

An effective physical measure to counter the spread of COVID-19 is the utilization of face masks in public settings [33]. **Figure 1** illustrates the varying transmission risks between an infected individual and an uninfected person. When an infected person does not wear a mask, the risk of transmitting the virus to an uninfected person is substantially high, as depicted in the first row. This risk diminishes to a moderate level if either of them wears a face mask (depicted in the second row). The lowest risk of

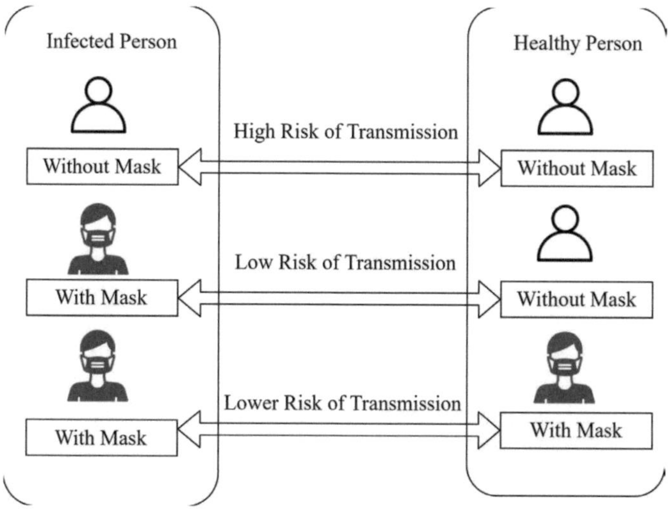

Figure 1.
Different risk of transmission between infected person (left column) and uninfected person (right column).

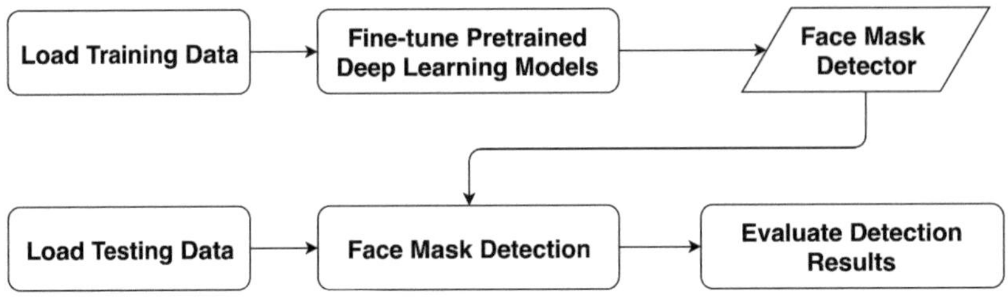

Figure 2.
The flow of building face mask detector.

infection occurs when both individuals are wearing masks [34], as depicted in the third row. Thus, wearing face masks effectively reduces the spread of COVID-19. However, ensuring universal adherence to mask-wearing mandates poses challenges. AI-driven face mask detection [35] has emerged as a technique to identify compliance with face mask requirements and can serve as a reminder for those not wearing masks. Developing a face mask detection system from scratch presents challenges due to the scarcity of labeled images. As a result, deep transfer learning [36] offers a promising solution to this predicament. This technique involves adapting pre-trained models to the task of face mask detection. The complete process of constructing face mask detection models through deep transfer learning is delineated in **Figure 2**.

During the model training phase, a limited set of annotated images is loaded as training data. These images are then used to fine-tune pre-trained deep learning models, ultimately creating a robust fake mask detector. In the testing phase, datasets with labeled ground truth are loaded. The face mask detector is applied to this data, and its performance is evaluated using predefined metrics. Subsequent sections delve into comprehensive details regarding COVID-19 screening methodologies and fake mask detection on edge devices, all supported by illustrative case studies.

The structure of this chapter is as follows: In Section 2, we provide an overview of previous research on AI-driven COVID-19 screening techniques and face mask detection. Moving to Section 3, we delve into specific case studies, presenting associated outcomes that highlight effective screening and face mask detection strategies. These endeavors leverage deep learning and edge devices to mitigate the spread of COVID-19. Finally, Section 4 encapsulates the conclusions and outlines avenues for future research.

2. Related work

2.1 COVID-19 screening *via* AI-enhanced image processing

The advancement of AI-driven image processing has played a pivotal role in significantly advancing COVID-19 screening techniques, encompassing both COVID-19 classification and COVID-19 segmentation. Considerable attention has been directed toward the classification of COVID-19 cases versus non-COVID-19 cases, with a focus on employing deep learning models. These models aim to effectively differentiate between COVID-19 patients and non-COVID-19 subjects, wherein the latter group includes individuals with common pneumonia and those without pneumonia. The diagram depicted in **Figure 3** illustrates the process of COVID-19 classification on a chest X-ray image. The COVID-19 classifier receives a chest X-ray image as

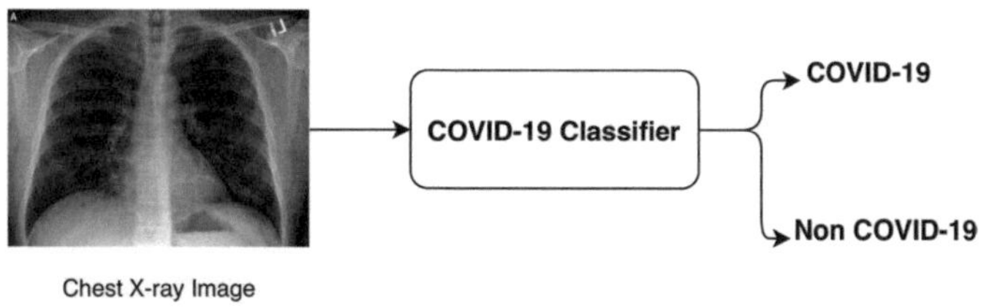

Figure 3.
A diagram of COVID-19 classification on chest X-ray images.

input and provides an output that classifies the image as either indicating COVID-19 or non-COVID-19 status.

Considerable research efforts have been directed toward COVID-19 classification. Chen et al. [27] pursued COVID-19 classification using segmented lesion patterns extracted *via* UNet++. Their dataset included diverse patient cases, such as COVID-19 patients, viral pneumonia patients, and non-pneumonia patients. Given the visual similarity between common pneumonias, particularly viral pneumonia, and COVID-19, distinguishing these conditions becomes crucial for effective clinical screening. To address this, a 2D CNN model was proposed, employing manually delineated region patches for classification between COVID-19 and typical viral pneumonia. Additionally, Wang et al. [37] combined segmentation information with a proposed 2D CNN model to classify COVID-19 cases by considering handcrafted features of relative infection distance from the lung's edge. Xu et al. [38] utilized candidate infection regions segmented by V-Net. They combined these region patches with handcrafted features representing the distance from the edge of the region to perform COVID-19 classification using a ResNet-18 model. Zheng et al. [24] employed U-Net for lung segmentation and utilized 3D CNNs to predict COVID-19 probabilities based on the segmented features. Their dataset consisted solely of chest CT images of COVID-19 and non-COVID-19 cases. Similarly, Jin et al. [29] introduced a UNet++ – based segmentation model to identify lesions and a ResNet50-based classification model for diagnosis. Their larger dataset encompassed chest CT images of 1136 cases, including 723 COVID-19 positives and 413 COVID-19 negatives. In another work, Jin et al. [39] employed a 2D Deeplab v1 model for lung segmentation and a 2D ResNet152 model for slice-based identification of positive COVID-19 cases. In summary, ongoing efforts in COVID-19 classification primarily focus on learning from significant volumes of medical images. However, the application of these techniques is hindered by the considerable data requirements for building effective classifiers.

The segmentation of COVID-19 cases is achieved through image semantic segmentation techniques bolstered by deep learning models such as U-Net and V-Net [40, 41]. An illustrative instance of image semantic segmentation on a chest X-ray image is presented in **Figure 4**. This approach enables the precise identification and isolation of COVID-19-affected regions within medical images.

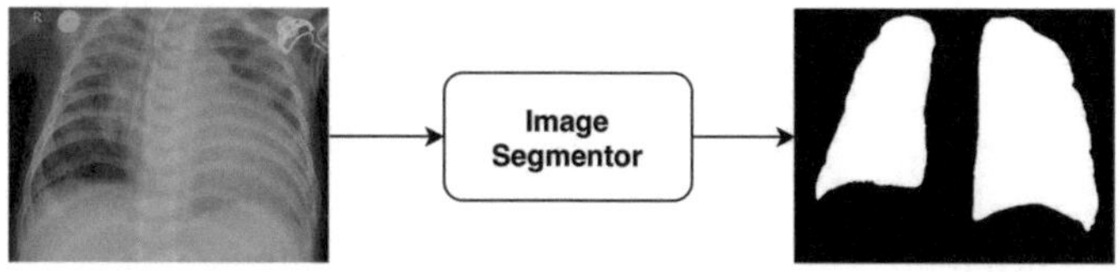

Figure 4.
An example of image semantic segmentation on a chest X-ray image.

The U-Net architecture is a powerful tool for segmenting both lung regions and lung lesions, facilitating the construction of effective image segmentation models [31]. U-Net, developed using a fully convolutional network [42], features a distinctive U-shaped design encompassing two symmetric paths: an encoding path and a decoding path. The layers at the corresponding levels in these two paths are interconnected through shortcut connections, fostering the acquisition of improved visual semantics and intricate contextual details. Zhou et al. [43] introduced UNet++, which inserts a nested convolutional structure between the encoding and decoding paths, further enhancing segmentation performance. In a similar vein, Milletari et al. [44] developed the V-Net, employing residual blocks as fundamental convolutional units and optimizing the network using a Dice loss. Furthermore, Shan et al. [28] devised VB-Net, enhancing segmentation efficiency by incorporating convolutional blocks with bottleneck blocks. Numerous variations of the U-Net architecture and its derivatives have been explored, yielding promising segmentation outcomes in COVID-19 diagnosis [27]. Advanced attention mechanisms are incorporated to identify the most discriminant features within deep learning models. Oktay et al. [45] introduced the Attention U-Net, capable of capturing intricate structures in medical images, rendering it suitable for segmenting lesions and lung nodules in COVID-19 applications. The integration of COVID-19 classification and segmentation empowers the implementation of multifaceted screening techniques across different levels.

2.2 Face mask detection

Face mask detection has been a subject of extensive research, and these efforts can be broadly categorized into two classes. The first approach treats it as an object detection task, where the goal is to localize the face mask area using bounding boxes. For instance, Mingjie Jiang et al. [46] proposed a one-stage face mask detector that employs a pre-trained ResNet for transfer learning and a feature pyramid network to extract semantic information. They introduced a novel context attention mechanism to enhance the detection of mask features. Similarly, Loey et al. [47] presented an object detection process using a combination of ResNet and YOLO V2. Another approach views face mask detection as an image classification problem [35]. Researchers in this category employed various convolutional neural networks (CNNs) such as MobileNet [48], Inception V3 [49], VGG-16 [50], and ResNet [51]. MobileNet, designed for edge devices, operates at high speed due to its smaller model size and complexity. Inception V3 utilizes factorizing convolutions to maintain robustness while reducing connections. VGG-16 explores the impact of depth on accuracy in large-scale image classification tasks.

In addition to training models from scratch, some researchers use pre-trained face detectors to extract faces, and then apply mask detection classification models on the detected faces [52, 53]. For example, Lippert et al. utilize OpenCV's pre-trained face detector and a VGG-16-based classifier for face mask detection. The deployment of machine learning models on resource-constrained devices has gained popularity, as executing models locally is often preferable to sending data to the cloud due to issues such as limited bandwidth and privacy concerns.

In summary, face mask detection has been tackled through two main avenues: object detection using bounding boxes and image classification using various CNN architectures. These diverse approaches aim to enhance the accuracy and efficiency of detecting face mask presence and adherence.

2.3 Edge device

Machine learning models are executed on small IoT or edge devices, which highlights the relevance of edge computing However, executing such models on these devices is not a straightforward task due to their computational intensity. Models must be lightweight or compatible with these devices to be feasible. NVIDIA Jetson TX2 and Nano are popular test devices. These NVIDIA devices serve as embedded AI computing solutions. The Jetson TX2 features an NVIDIA Pascal GPU with 8 GB of memory, while the Nano is equipped with a Maxwell GPU and 4 GB of memory. Both devices are well-suited for computer vision and machine learning tasks. **Figure 5** depicts the devices used in this chapter, and **Table 1** provides a detailed comparison between the two devices.

Figure 5.
NVIDIA Jetson Nano (left) and NVIDIA Jetson TX2 (right).

	Jetson TX2	Jetson Nano
CPU	Dual-Core NVIDIA Denver 2 64-Bit CPU	Quad-core ARM Cortex-A57 MPCore processor
GPU	256-core Pascal @1300 MHz	NVIDIA Maxwell architecture with 128 NVIDIA CUDA® cores
Memory	8GB 128-bit LPDDR4	4 GB 64-bit LPDDR4
Storage	32GB eMMC 5.1	16 GB eMMC 5.1

Table 1.
Comparison between NVIDIA Jetson TX2 and NVIDIA Jetson Nano.

3. Case study

3.1 Case study on COVID-19 classification

In the context of this chapter, chest X-ray images were selected for both model development and validation. It was driven by the cost-effectiveness and greater accessibility of chest X-rays compared to CT scans, particularly in communities with limited medical resources. Additionally, chest X-rays offer a swift imaging solution, making them particularly attractive for large-scale patient screening during the COVID-19 pandemic. Therefore, utilizing chest radiography for screening COVID-19 patients is considered a practical, efficient, and rapid approach [54, 55]. For validation purposes, this chapter utilized a comprehensive chest X-ray dataset "*COVIDx*" [23]. This dataset comprises a vast collection of 18,543 chest radiography images from 13,725 unique cases. It is important to note that when evaluating the distribution of classes between the training and testing datasets, a significant dissimilarity in class distribution becomes apparent.

This chapter employs two types of deep learning models: Convolutional Neural Networks (CNNs) and ResNet. CNNs have played a significant role in advancing various visual processing tasks like image classification [56], object detection and tracking [57, 58], and semantic segmentation [59]. The progress of CNNs has been facilitated by large datasets like ImageNet [56] and YouTube-BoundingBoxes [60], which provide ample training data for building large-scale models. The general architecture of CNN for image classification is depicted in **Figure 6**. SOTA CNN architectures such as AlexNet [56], VGG [50], and GoogleNet [61] have propelled advancements in image classification. These architectures leverage millions of annotated samples from large datasets to successfully estimate appropriate parameters. Furthermore, CNNs have been enhanced by combining them with other deep learning models. For example, Wang et al. [62] combined CNN with Recurrent Neural Networks (RNNs) for multi-label image classification. Additionally, CNNs combined with autoencoders [63, 64] have demonstrated effectiveness in tasks like face detection. In this chapter, three small CNNs were trained and tested on a subset of the COVIDx dataset to demonstrate a proof-of-concept experiment.

ResNet [51] is an artificial neural network architecture inspired by the structure of pyramidal cells in the cerebral cortex. It introduces skip connections, or shortcuts, which allow the network to bypass certain layers. The concept behind ResNet is that training a network to learn a residual mapping is simpler than training it to directly learn the underlying mapping. This is achieved using residual blocks, as depicted in **Figure 7**. A crucial modification in ResNet compared to a standard CNN is the "skip connection"

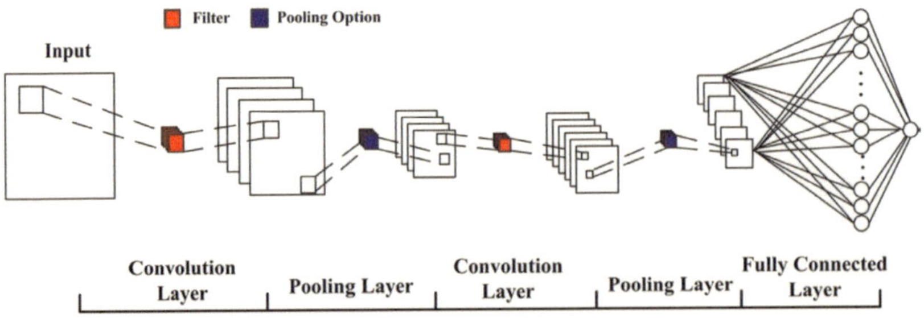

Figure 6.
Convolutional neural network architecture for image classification.

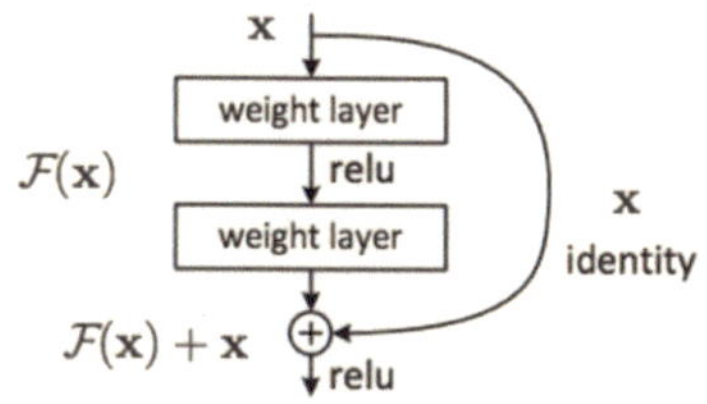

Figure 7.
General architecture of the residual block of ResNet [51].

for identity mapping. This identity mapping has no parameters; it simply adds the output from the previous layer to the next layer. However, the dimensions of x and $F(x)$ might differ. Since convolutions usually reduce spatial resolution, e.g., a 3×3 convolution on a 32×32 image results in a 30×30 image, the identity mapping is expanded using a linear projection W to match the channels of the residual. This allows the input x and $F(x)$ to be combined as input to the subsequent layer. Given the effectiveness of ResNet, various ResNet architectures will be employed in this chapter to screen COVID-19 using the complete COVIDx dataset.

This chapter examines the proposed models from two distinct viewpoints. The first perspective involves assessing their capability to effectively identify COVID-19 cases from a limited dataset, using compact CNNs. The second perspective entails investigating whether these models can utilize the ResNet architecture to identify COVID-19 cases within an extensive dataset, all without relying on transfer learning techniques. For the small dataset scenario, a subset of 350 images was extracted from the original COVIDx dataset [23] for training and testing. Among these, 300 images were allocated for training, while the remaining 50 were reserved for testing. Three small CNNs were employed for this evaluation. The obtained training and testing accuracies for these three models are depicted in **Figure 8**. The results indicate that the initial shallow CNN (referred to as CNN1) encountered issues with under-fitting. However, as additional layers were incorporated to extract more intricate features, CNN3 exhibited superior accuracy performance.

In the context of a large dataset, all images encompassed within the COVIDx dataset were harnessed to establish a classifier for COVID-19 identification. The initial step encompassed data preprocessing, involving the compression of images. The

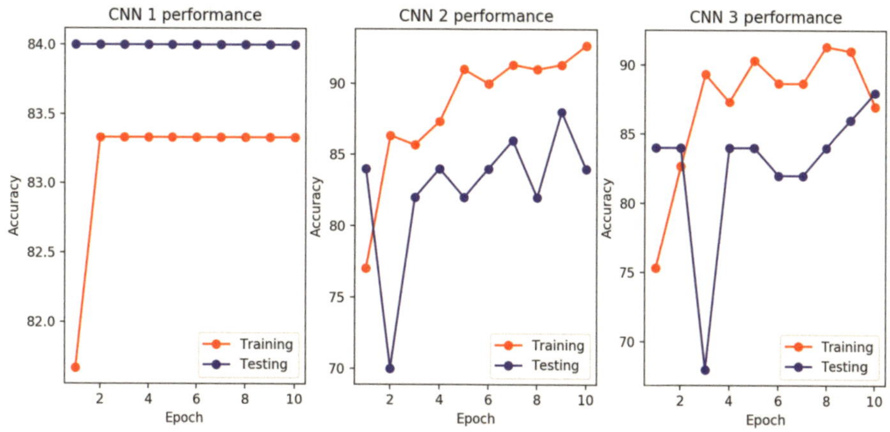

Figure 8.
Performance on training and testing accuracy for three small CNNs.

original X-ray images within the dataset measured 1024 × 1024 × 3 in dimensions. To expedite the training process, the image size was compressed to 64 × 64 × 3. Subsequently, training was initiated using this preprocessed dataset. The outcomes of the training and testing phases are depicted in **Figure 9**.

The findings unveil the existence of an optimal configuration, specifically ResNet-34, which outperforms alternative models. Beyond ResNet-34, a consistent better performance becomes apparent as the number of ResNet layers increases to 152. This can be attributed to the phenomenon where, with the progressive augmentation of layers, the models tend to overfit while the available data remains inadequate to effectively train the model. It is noteworthy that although the training time slightly extended with the inclusion of additional layers in the ResNet models, the performance gains were limited. Furthermore, the results underscore the accomplishment of this chapter in achieving commendable performance by training ResNet models from scratch, devoid of reliance on transfer learning techniques.

While supervised deep learning demonstrates impressive performance in classifying COVID-19 images, its practical application is hindered by the need for a substantial volume of annotated medical images for training. Given the limitations in available COVID-19-related data resources and the significant costs associated with labeling medical images, this approach becomes less feasible, further exacerbated by labeling inaccuracies that may arise [65]. To address this challenge, the focus has shifted toward semi-supervised deep learning, which has garnered considerable attention due to its capacity to enhance model generalization by leveraging both labeled and unlabeled data [66–69]. This paradigm involves training deep neural networks through the simultaneous optimization of a standard supervised classification loss on labeled samples and an unsupervised loss on unlabeled data [67, 69]. Semi-supervised learning models aim to amplify the information derived from unlabeled data [70] or impose regularization on the network to enforce smoother and more consistent classification boundaries [68].

In the realm of COVID-19 research, particularly in tasks like COVID-19 image classification and image segmentation, semi-supervised learning has emerged as a solution to mitigate the scarcity of labeled data [71–76]. However, within the domain of COVID-19 image classification, the studies conducted by Zhou et al. [76], Calderon et al. [72], and Paticchio et al. [74] have not thoroughly examined model performance

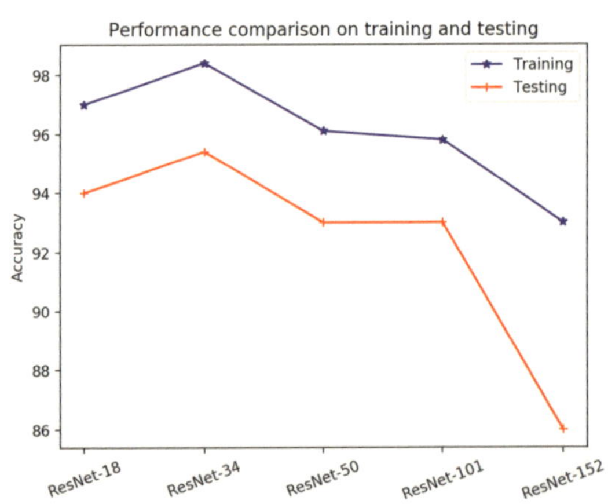

Figure 9.
Performance comparison on training and testing.

using large-scale X-ray image datasets such as *COVIDx* [23]. Furthermore, they have not conducted comprehensive comparisons against state-of-the-art methods, particularly in scenarios where labeled data is severely limited, constituting less than 10% of the dataset. In response, this chapter introduces a semi-supervised deep learning model for COVID-19 image classification, systematically evaluating its performance on the *COVIDx* dataset [23].

Using the ResNet architecture, this chapter devised a two-path semi-supervised learning ResNet (referred to as SSResNet), which comprises three key components: a shared ResNet, a supervised ResNet, and an unsupervised ResNet. These two paths are formed by coupling the shared ResNet with either a supervised ResNet or an unsupervised ResNet. Both labeled and unlabeled data are leveraged in the computation of the unsupervised loss, utilizing the mean squared error loss (MSEL). Conversely, only labeled data contributes to the calculation of the supervised loss, employing the cross-entropy loss (CEL). To counterbalance data imbalance, a weighted cross-entropy loss (WCEL) was ingeniously crafted, assigning greater weight to the COVID-19 class. The primary objective of minimizing MSEL lies in enhancing image representation, while the reduction of WCEL aims to boost classification performance. For a comprehensive outline of the methodology, refer to Algorithm 1. The efficacy of the proposed model was thoroughly assessed using the extensive X-ray image dataset *COVIDx*. The experimental findings distinctly establish that the proposed model excels in the realm of COVID-19 image classification. Notably, even when trained with a notably limited quantity of labeled X-ray images, the model showcases remarkable performance.

Algorithm 1. Learning of Semi-supervised ResNet (SSResNet).

Require: training sample x_i, the set of training samples S, label y_i for x_i $(i \in S)$

1. **for** t in [1, num epochs] **do**
2. **for** each minibatch B **do**
3. $z_{i \in B} \leftarrow f_{\theta_{shared}}(x_{i \in B})$ ▷ shared representation
4. $z^{sup}_{i \in B} \leftarrow f_{\theta_{sup}}(z_{i \in B})$ ▷ supervised representation
5. $z^{unsup}_{i \in B} \leftarrow f_{\theta_{unsup}}(z_{i \in B})$ ▷ unsupervised representation
6. $l^{WCEL}_{i \in B} \leftarrow -\frac{1}{|B|}\sum_{i \in B \cap S} log\phi\left(z^{sup}_i\right)[y_i w_i]$ ▷ supervised loss component
7. $l^{MSEL}_{i \in B} \leftarrow \frac{1}{C|B|}\sum_{i \in B}\left\|z^{sup}_i - z^{unsup}_i\right\|^2$ ▷ unsupervised loss component
8. $Loss \leftarrow l^{WCEL}_{i \in B} + \lambda \times l^{MSEL}_{i \in B}$ ▷ total loss
9. update $\theta_{shared}, \theta_{sup}, \theta_{unsup}$ using optimizer, e.g., ADAM

return $\theta_{shared}, \theta_{sup}, \theta_{unsup}$

Upon analyzing the class distribution disparities between our training and testing datasets, a noteworthy distinction came to light. Consequently, we undertook the task of reconstructing the data structure. This involved partitioning the dataset into distinct training and testing subsets, ensuring that their class distributions closely aligned. Specifically, 70% of the data was allocated for the training dataset, while the remaining 30% constituted the testing dataset. For a comprehensive breakdown of the reconstructed dataset, including sample distribution details, please refer to **Table 2**.

It is evident that the distribution of samples in our dataset is highly skewed, particularly in relation to the COVID-19 class. This imbalance presents a significant hurdle in achieving a high-performing classifier. To surmount this issue, our proposed model employs a weighted cross-entropy loss. This innovative approach involves according greater weight to the minority class (COVID-19) throughout the training process. For a comprehensive understanding of this technique, please refer to section two.

In the experiment, the key hyper parameters for training the proposed model are: Minibatch size: 256, Number of epoch: 50, Optimizer: Adam optimizer, and Initial Learning rate: 0.1. They are determined by trial and error. Moreover, the details of the model architecture are illustrated in **Table 3**. We employ COVID-Net[1] as a baseline supervised model to present the state-of-the-art performance of COVID-19 image classification for comparison. Furthermore, we compared the proposed model with SRC-MT that is the state-of-the-art semi-supervised learning since it outperformed Π model and mean teacher model in the area of medical image classification.

Table 4 showcases a comprehensive comparison of the classification performance between SRC-MT and the newly introduced model (SSResNet). On the whole, the overall accuracies attained by SRC-MT surpass those achieved by the proposed SSResNet. Nonetheless, an interesting observation emerges when considering scenarios where merely 5% of labeled samples were utilized for training. In this specific scenario, the MacroF metric of SSResNet surpasses that of SRC-MT. This outcome indicates that the proposed model exhibits greater efficacy in identifying COVID-19 samples. Essentially, the utilization of the unsupervised path in SSResNet appears to remarkably enhance data representation, thereby leading to a more pronounced improvement in COVID-19 classification performance compared to SRC-MT.

Furthermore, this study delved into a granular examination of the performance for each class, elucidating the outcomes through the employment of confusion matrices, as depicted in **Figure 10**. A noteworthy observation emerges from this analysis: The proposed model outperforms SMC-TC in terms of recognizing COVID-19 cases. This observation underscores the capability of SSResNets to glean more effective features from unlabeled data, resulting in a heightened ability to discern COVID-19 samples. Notably, as the ratios of labeled data are incrementally augmented, there is a marked improvement in the accuracy of COVID-19 recognition. This underscores the proposition that the inclusion of an unsupervised path serves to elevate image representations, consequently bolstering classification performance. In essence, the integration of unlabeled data distinctly contributes to a substantial enhancement in COVID-19 classification performance, primarily attributable to the amplification of image representations facilitated by the unsupervised path within the SSResNet.

---

[1] https://github.com/lindawangg/COVID-Net

Dataset	Normal	Pneumonia	COVID-19	Total
Training	6195	6708	75	12,978
Testing	2656	2876	33	5565
Total	8851	9584	108	18,543

Table 2.
Sample distribution in different classes for training and testing datasets.

Name	Description
Input	Medical Images
Shared ResNet	one convolutional layer, 2 residual block
	batch normalization, one pooling layer
Supervised ResNet	one convolutional layer, 2 residual block
	batch normalization, one pooling layer
Unsupervised ResNet	one convolutional layer, 2 residual block
	batch normalization, one pooling layer
Output	image class $\phi(z^{sup})$ and
	a new representation z^{unsup}

Table 3.
The proposed network architecture.

Semi-supervised model	Accuracy (%)	MacroP (%)	MacroR (%)	MacroF (%)
SRC-MT (5%)	90.67	61.08	60.75	60.59
SRC-MT (7%)	89.82	89.92	74.13	78.95
SRC-MT (9%)	92.79	93.61	79.15	84.15
Our model	Accuracy (%)	MacroP (%)	MacroR (%)	MacroF (%)
SSResNet (5%)	84.95	61.18	66.76	62.41
SSResNet (7%)	84.21	63.67	67.85	62.83
SSResNet (9%)	81.79	59.34	70.99	59.19

Table 4.
Comparing performance between SRC-MT and proposed model (semi-supervised ResNet (SSResNet)).

3.2 Face mask detection

This chapter also introduces the process of constructing a mobile model designed for face mask detection on edge devices. The primary goal of this endeavor is to develop an intelligent Internet of Things (IoT) device capable of performing real-time video processing to ascertain whether an individual is wearing a face mask in public settings. This technology finds practical application in scenarios such as enforcing mask-wearing within facilities or buildings. In this context, a smart IoT camera functions as a vigilant observer, detecting instances where individuals are not complying with the mask mandate and triggering alarms as necessary. To achieve this, mobile devices like FPGAs, integrated into the camera system, execute the face mask

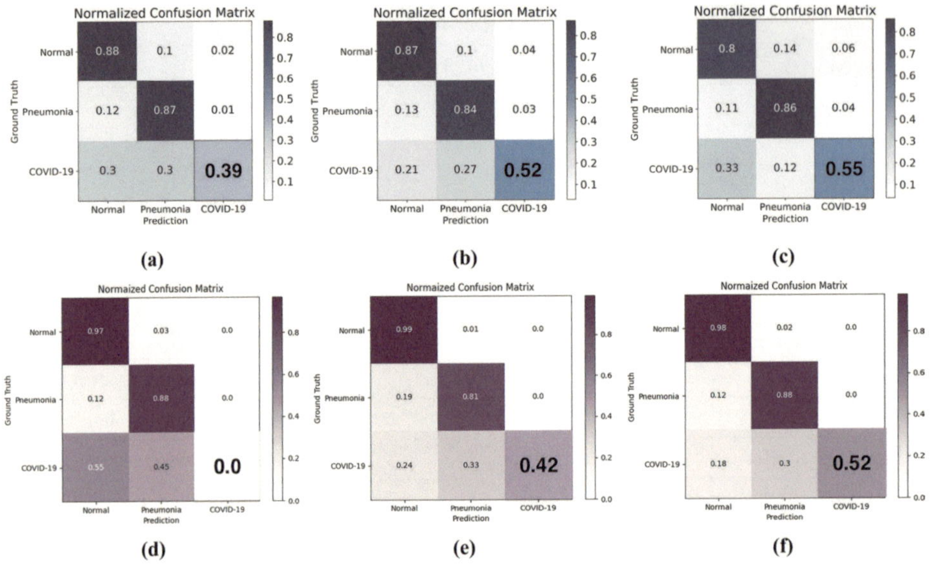

Figure 10.
Comparison of confusion matrix generated with SRC-MT and SSResNets trained on different ratios of labeled data.

detection models. For a visual representation of this concept, refer to **Figure 11**, which presents an illustrative diagram depicting the potential of employing face mask detection to control access to a door.

Instead of embarking on the construction of models from scratch, the approach taken involves the utilization of pre-trained models that have undergone training using more extensive datasets across broader classes. This pre-trained ensemble comprises four distinct convolutional neural networks (CNNs): MobileNet V2, Inception V3, VGG 16, and ResNet 50. Leveraging transfer learning, as opposed to constructing models from scratch, often leads to superior performance. These pre-trained models have been honed on the comprehensive ImageNet dataset [56]. The salient attributes extracted from these pre-trained models serve as the foundation, conveyed to a novel classifier positioned at the terminal end of the network. Subsequently, a mask detection classifier is trained on top of the pre-trained model. A pivotal distinction discernible among these models is their input size. Inception V3 adopts an image size of $299 \times 299 \times 3$, while the other three models employ a size of $224 \times 224 \times 3$. The efficacy of these models is rigorously verified through inference runs on mobile devices, encompassing the NVIDIA Jetson TX2 and NVIDIA Jetson Nano platforms.

Dataset and experiment setup are presented below.

- Dataset: We utilized publicly available datasets[2], which consist of 3890 images categorized into two classes: with face and without face. Among these, 1916 images featured individuals wearing masks, while 1930 images depicted individuals without masks, after removing redundant entries. This dataset was deliberately structured to maintain a balanced distribution for our classifier. However, it does present some exceptional cases, such as images containing multiple faces or instances where faces are partially occluded by other body parts.

[2] https://github.com/chandrikadeb7/Face-Mask-Detection

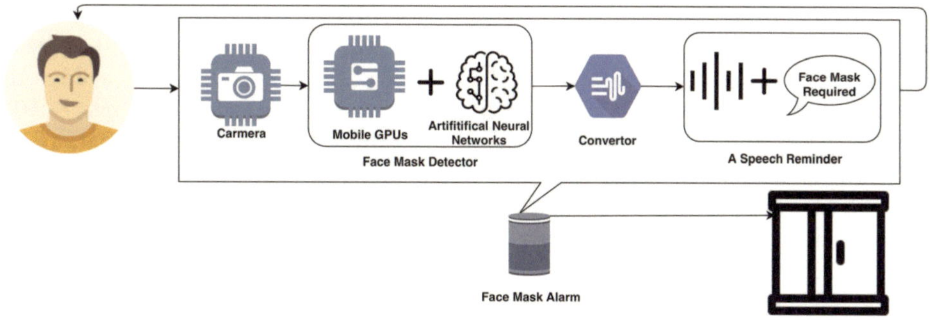

Figure 11.
A diagram of a face mask alarm system. A face mask alarm mounted on the door consists of a camera, a face mask detector, and a convertor. The camera will send the personal image to the face mask detector consisting of mobile GPUs and artificial neural networks. Then, the detector will detect the face mask on the image by running the neural networks on the mobile GPUs and sent the detection result. Finally, the convertor will show a speech reminder such as "Face Mask Required" back to the person if the detection result indicates there is no face mask on the face.

- Experiment setup: A learning rate of 0.0001 was employed, while the experiment consistently utilized a batch size of 10. This modest batch size was deliberately selected to facilitate extended training while operating within memory limitations. Furthermore, the experimentation spanned 100 epochs. The loss function of choice was binary cross-entropy.

The experiment involves the utilization of publicly available datasets, sourced from https://github.com/chandrikadeb7/Face-Mask-Detection. This dataset comprises a total of 3890 images that include both images with faces and images without faces. Within this collection, there are 1916 images depicting individuals wearing masks and 1930 images without masks. This dataset design maintains a balanced distribution, which is optimal for classification. It is important to note that certain exceptions are present, such as images containing multiple faces or faces partially covered by other body parts. In configuring the models, a learning rate of 0.0001 is employed. Throughout the experiment, a batch size of 10 is consistently used. This choice is rooted in the goal of facilitating more extensive training while optimizing memory usage. The training process spans 100 epochs, and for the loss function, the binary cross-entropy is selected.

In the model implementation, a diverse set of tools come into play. TensorFlow [77], a widely adopted open-source software library, forms the core framework for machine learning applications, supporting operations across various processing units including CPUs, GPUs, and TPUs. The flexibility it offers, in terms of both architectural design and working with pre-trained models, contributes to its popularity. Keras [78], which functions atop TensorFlow, is another critical component. Designed for expedited execution of deep learning models, Keras enhances the speed of development. For this research, Keras is employed to leverage pre-trained deep learning models, which are then fine-tuned to create the face mask detection classifier through transfer learning.

This study delved into the robustness of the models by subjecting them to training on limited sample sizes. Traditionally, working with small training datasets results in diminished training and testing performance. However, given the unique context of face mask detection in the context of the COVID-19 outbreak, acquiring extensive data for training and testing becomes a challenge. Thus, developing models for face mask detection that excel with small training data becomes pivotal for creating effective applications. Additionally, these models must be locally deployable and capable of achieving tasks at an optimal speed.

To address these requirements, various ratios of training data were employed to assess model performance. The outcomes, presented in **Table 5**, were obtained by running these models on mobile NVIDIA GPUs. The experimental results yield noteworthy insights. MobileNet V2 emerges as the swiftest model, processing nearly 40 frames per second (FPS) on the Jetson TX2 platform. On the other hand, VGG 16 attains the highest accuracy among the models. Notably, when trained using just 1% of the available training data, Inception demonstrates superior performance.

Observations indicate that training accuracy surpasses 90% for each model, indicative of overfitting given the markedly lower testing scores. As the proportion of training data is gradually increased to 5, 10, and 20%, the performance of all models exhibits enhancement. This suggests that augmenting the dataset samples contributes to alleviating overfitting issues. VGG secures the highest accuracy, and while training on 20% of the data, MobileNet attains the lowest accuracy. ResNet and Inception showcase comparable performance levels throughout this experimentation.

Additionally, it is observed in **Table 6** that the pretrained models performed better on recognizing images containing masks regarding the precision, recall, and Fscore. Additionally, these pretrained models can achieve promising performance in terms of Fscore.

In summary, we leveraged pre-trained models such as MobileNet V2, ResNet 50, Inception V3, and VGG 16. These models were chosen due to their established performance in various applications. When considering model complexity, MobileNet V2 stands out for its efficiency, boasting the lowest complexity among the aforementioned

Models	**Training**				**Testing on TX2**			**Testing on Nano**		
	R_{tr}	N_{tr}	Loss	Training accuracy	Testing accuracy	Inference time (ms)	FPS	Testing accuracy	Inference time (ms)	FPS
MobileNet V2	1%	30	0.107	0.9776	0.6957	26.97	37.27	0.6926	42.40	23.58
	5%	150	0.095	0.9733	0.7685	25.10	39.86	0.7659	42.00	23.80
	10%	300	0.114	0.9622	0.7842	25.10	39.87	0.7801	42.90	23.31
	20%	601	0.136	0.9490	0.7735	25.53	39.14	0.7699	43.03	23.24
ResNet 50	1%	30	0.071	0.9666	0.5000	70.90	14.11	0.7433	195.05	5.13
	5%	150	0.050	0.9845	0.7976	71.00	14.09	0.8023	176.45	5.67
	10%	300	0.035	0.9900	0.7976	70.13	14.25	0.7962	173.05	5.78
	20%	601	0.027	0.9927	0.8651	71.00	14.08	0.8599	173.65	5.76
Inception V3	1%	30	0.167	0.9556	0.8275	89.27	11.21	0.8275	187.00	5.35
	5%	150	0.136	0.9578	0.8704	87.80	11.41	0.8713	188.53	5.31
	10%	300	0.136	0.9445	0.9066	92.33	10.85	0.9057	905.72	5.22
	20%	601	0.167	0.9379	0.8972	89.57	11.17	0.8977	192.77	5.19
VGG 16	1%	30	0.089	0.9667	0.6857	2138.77	0.47	0.6761	239.37	4.18
	5%	150	0.006	0.9978	0.9231	2139.60	0.47	0.9173	245.23	4.08
	10%	300	0.010	0.9956	0.9370	2136.00	0.47	0.9325	236.17	4.23
	20%	601	0.006	0.9994	0.9607	2139.57	0.47	0.9580	237.70	4.21

Table 5.
Performance comparison on face mask detection on Jetson TX2 and Jetson Nano. FPS refers to the number of images processed per second. R_{tr} and N_{tr} denote the ratio of training data and the number of training images.

Models	Class	Precision	Recall	Fscore
MobileNet V2	With Mask	0.99	0.83	0.90
	Without Mask	0.86	0.99	0.92
ResNet 50	With Mask	1.00	0.96	0.98
	Without Mask	0.96	1.00	0.98
Inception V3	With Mask	1.00	0.98	0.99
	Without Mask	0.98	1.00	0.99
VGG 16	With Mask	1.00	1.00	1.00
	Without Mask	1.00	1.00	1.00

Table 6.
Comparison of precision, recall, and Fscore.

models. On the other hand, VGG-16 adheres to a classical architecture characterized by a substantial number of parameters, leading to an overall higher complexity. InceptionV3, with its emphasis on capturing multi-scale features, adopts a more intricate architecture than MobileNet V2. In the context of speed and accuracy, as showcased in the performance comparison table labeled "Performance comparison on face mask detection on Jetson TX2 and Jetson Nano," MobileNet V2 outperforms the others in terms of speed. In contrast, VGG-16 exhibits the slowest processing speed, aligning with the principle that more complex models tend to operate at a slower pace. Furthermore, it is noteworthy that VGG-16 achieves optimal performance relative to the other models, driven by its expansive model capacity. This insight underscores the trade-off between complexity and performance, where VGG-16's higher capacity contributes to its superior performance despite the trade-off in processing speed.

3.3 Challenges to real-world applications

Deploying COVID screening for real-world applications still presents several challenges:

- COVID screening: To be effective in real-world scenarios, COVID screening requires a streamlined and efficient process. However, existing techniques, particularly deep learning-based X-ray COVID detection, involve intricate steps, including chest X-ray tests and subsequent COVID classification. Moreover, the quality of data obtained from chest X-ray tests might not be optimal for accurate COVID classification. Additionally, the reliance on high-performance computing hardware for deep learning-based classification hinders its applicability in underdeveloped regions.

- Face Mask Detection: It encounters several obstacles that hinder its real-world applications. Firstly, the high accuracy achieved in controlled experiments might not translate directly to real-world scenarios due to significant differences in image backgrounds, object sizes and positions, and object overlaps. Secondly, face mask detection must work across various environments, including both indoor and outdoor settings. Adapting to outdoor environments requires techniques capable of handling diverse weather conditions and backgrounds, which poses difficulties for current methodologies. Lastly, users prefer seamless face mask detection that does not disrupt their daily activities. This demands the development of techniques that can perform detection without requiring users to actively interact with a camera.

In both cases, addressing these challenges requires innovative solutions that can simplify processes, adapt to diverse conditions, and accommodate user preferences without compromising performance or accuracy.

4. Conclusion

The deployment of technologies such as COVID-19 screening and counterfeit mask detection plays a crucial role in curbing the spread of the COVID-19 virus. This chapter has introduced advanced AI-driven methods designed to enhance both COVID-19 screening and the identification of face masks. Specifically, the utilization of deep learning for COVID-19 classification has demonstrated significant potential in the creation of effective screening tools. Moreover, the application of deep learning for face mask detection on mobile devices has exhibited promising results. In the pursuit of developing these innovative techniques, valuable insights are gained that can be applied in future pandemic situations. In the future, potential work includes: (1) Overfitting is characterized by machine learning models achieving impressive performance during the training phase, but faltering when tested on new data. To ensure the efficacy of large models for COVID screening and face mask detection, it is imperative to address this concern. In our forthcoming work, we aim to tackle this issue through the implementation of data augmentation techniques, leveraging expansive pre-trained image models. More specifically, these large pre-trained image models hold the capacity to generate synthetic data for both COVID screening and face mask detection. This synthetic data will encompass varying backgrounds, enriching the diversity of the training dataset and reducing overfitting tendencies. Additionally, we intend to incorporate human feedback as an integral part of the loop. This iterative process will involve human assessment to gauge the quality of the generated data, fostering continuous improvements in the data augmentation strategy. This combined approach of data augmentation and human feedback holds significant promise in enhancing the generalization capabilities of the models, thereby enabling robust and high-performing COVID screening and face mask detection systems in real-world scenarios; (2) AI ethics continue to be a significant concern across various applications, encompassing issues such as privacy breaches and biased data. In our future efforts, we are dedicated to addressing these ethical challenges head-on. Our strategy involves the application of privacy-preserving techniques to safeguard user privacy, coupled with measures to mitigate biased data. To protect user privacy, we are poised to employ our pioneering privacy-preserving edge intelligent computing framework. This entails training autoencoders in an unsupervised manner on individual edge devices. Subsequently, the latent vectors derived from these autoencoders are transmitted to the edge server for classifier training. This approach effectively reduces communication overhead while safeguarding end-users' sensitive data from exposure. In tackling biased data concerns, our plan is to integrate fair pre-processing techniques from AIF360, an AI fairness toolkit.[3] These techniques will be strategically applied during the data collection phase

[3] https://github.com/Trusted-AI/AIF360

for both COVID screening and face mask detection. By doing so, we aspire to counteract biases that may emerge in the data, ensuring equitable and unbiased outcomes. By proactively addressing privacy and bias concerns through cutting-edge privacy-preserving frameworks and fairness techniques, we aim to develop AI solutions that not only excel in performance but also uphold the highest standards of ethical conduct.

Acknowledgements

This research work is supported by the US NSF award 2018945, 2205891, and 2302469.

Conflict of interest

The authors declare no conflict of interest.

Author details

Xishuang Dong[1*†], Lucy Nwosu[1†], Sheikh Rufsan Reza[2†] and Xiangfang Li[2†]

1 Department of Electrical and Computer Engineering and the Center for Computational Systems Biology, Prairie View A&M University, Texas A&M University System, Prairie View, USA

2 Department of Electrical and Computer Engineering and the Center of Excellence in Research and Education for Big military Data Intelligence, Prairie View A&M University, Texas A&M University System, Prairie View, USA

*Address all correspondence to: xidong@pvamu.edu

† These authors contributed equally.

References

[1] Organization W.H et al. Statement on the Second Meeting of the International Health Regulations Emergency Committee Regarding the Outbreak of Novel Coronavirus (2019-nCoV). Geneva, Switzerland; 2020

[2] Organization W.H et al. WHO Director-general's Opening Remarks at the Media Briefing on COVID-19-11 March 2020. Geneva: Switzerland; 2020

[3] CDC. Testing for COVID-19. Available from: https://www.cdc.gov/coronavirus/2019-ncov/symptoms-testing/testing.html. [Accessed: April 19 2020]

[4] NY Times. Germany got testing right, what can we learn?. Available from: https://www.nytimes.com/2020/04/28/opinion/coronavirus-testing-united-states.html. [Accessed: April 29 2020]

[5] Ai T, Yang Z, Hou H, Zhan C, Chen C, Lv W, et al. Correlation of chest CT and RT-PCR testing for coronavirus disease 2019 (COVID-19) in China: A report of 1014 cases. Radiology. 2020; **296**(2):32-40

[6] Fang Y, Zhang H, Xie J, Lin M, Ying L, Pang P, et al. Sensitivity of chest CT for COVID-19: Comparison to RT-PCR. Radiology. 2020;**296**(2):115-117

[7] Liang T et al. Handbook of COVID-19 Prevention and Treatment. Zhejiang: Zhejiang University School of Medicine; 2020

[8] Kanne JP. Chest CT findings in 2019 novel coronavirus (2019-nCoV) infections from Wuhan, China: Key points for the radiologist. Radiology. 2020;**295**(1):16-17

[9] Bernheim A, Mei X, Huang M, Yang Y, Fayad ZA, Zhang N, et al. Chest CT findings in coronavirus disease-19 (COVID-19): Relationship to duration of infection. Radiology. 2020;**295**(3): 685-691

[10] Narin A, Kaya C, Pamuk Z. Automatic detection of coronavirus disease (COVID-19) using X-ray images and deep convolutional neural networks. Pattern Analysis and Applications. 2021; **24**(3):1207-1220

[11] Bullock J, Luccioni A, Pham KH, Lam CSN, Luengo-Oroz M. Mapping the landscape of artificial intelligence applications against COVID-19. Journal of Artificial Intelligence Research. 2020; **69**:807-845

[12] Bengio Y, Courville A, Vincent P. Representation learning: A review and new perspectives. IEEE Transactions on Pattern Analysis and Machine Intelligence. 2013;**35**(8): 1798-1828

[13] Dahl GE, Ranzato MA, Mohamed AR, Hinton G. Phone recognition with the mean-covariance restricted Boltzmann Machine. In: Proceedings of the 23rd International Conference on Neural Information Processing Systems. New Orleans Ernest N. Morial Convention Center; Dec 2010. pp. 469-477

[14] Deng L, Seltzer ML, Yu D, Acero A, Mohamed AR, Hinton G. Binary coding of speech spectrograms using a deep auto-encoder. In: Eleventh Annual Conference of the International Speech Communication Association. 2010. pp. 1692-1695

[15] Yu D, Seide F, Li G. Conversational speech transcription using context-dependent deep neural networks. In: Proceedings of the 29th International

Conference on Machine Learning. Edinburgh, Scotland. Jun 2021. pp. 1-2

[16] Hinton G, Deng L, Yu D, Dahl GE, Mohamed AR, Jaitly N, et al. Deep neural networks for acoustic modeling in speech recognition: The shared views of four research groups. IEEE Signal Processing Magazine. 2012;**29**(6):82-97

[17] Hinton GE, Osindero S, Teh YW. A fast learning algorithm for deep belief nets. Neural Computation. 2006;**18**(7): 1527-1554. DOI: 10.1162/neco.2006.18.7. 1527. URL doi:10.1162/neco.2006.18. 7.1527

[18] Glorot X, Bordes A, Bengio Y. Deep sparse rectifier neural networks. Aistats. 2011;**15**:275

[19] Krizhevsky A, Sutskever I, Hinton GE. Imagenet classification with deep convolutional neural networks. In: Advances in Neural Information Processing Systems. Lake Tahoe, Nevada, USA; 2012. pp. 1087-1105

[20] Ghoshal B, Tucker A. Estimating uncertainty and interpretability in deep learning for coronavirus (COVID-19) detection. 2020. arXiv preprint arXiv: 2003.10769

[21] Cohen JP, Morrison P, Dao L. COVID-19 image data collection. 2020. arXiv preprint arXiv:2003.11597

[22] Zhang J, Xie Y, Li Y, Shen C, Xia Y. COVID-19 screening on chest X-ray images using deep learning based anomaly detection. 2020. arXiv preprint arXiv:2003.12338

[23] Wang L, Wong A. COVID-Net: A tailored deep convolutional neural network design for detection of COVID-19 cases from chest radiography images. arXiv preprint arXiv:2003. 09871. 2020

[24] Zheng C, Deng X, Fu Q, Zhou Q, Feng J, Ma H, et al. Deep learning-based detection for COVID-19 from chest CT using weak label. MedRxiv. 2020

[25] Huang L, Han R, Ai T, Yu P, Kang H, Tao Q, et al. Serial quantitative chest CT assessment of COVID-19: Deep-learning approach. Radiology: Cardiothoracic Imaging. 2020;**2**(2):e200075

[26] Gozes O, Frid-Adar M, Greenspan H, Browning PD, Zhang H, Ji W, et al. Rapid AI development cycle for the coronavirus (COVID-19) pandemic: Initial results for automated detection and patient monitoring using deep learning CT image analysis. 2020. arXiv preprint arXiv:2003.05037

[27] Chen J, Wu L, Zhang J, Zhang L, Gong D, Zhao Y, et al. Deep learning-based model for detecting 2019 novel coronavirus pneumonia on high-resolution computed tomography: A prospective study. MedRxiv. 2020

[28] Shan F, Gao Y, Wang J, Shi W, Shi N, Han M, et al. Lung infection quantification of COVID-19 in CT images with deep learning. arXiv preprint arXiv:2003.04655. 2020

[29] Jin S, Wang B, Xu H, Luo C, Wei L, Zhao W. et al. AI-assisted CT imaging analysis for COVID-19 screening: Building and deploying a medical AI system in four weeks. MedRxiv. 2020

[30] Tang L, Zhang X, Wang Y, Zeng X. Severe COVID-19 pneumonia: Assessing inflammation burden with volume-rendered chest CT. Radiology: Cardiothoracic Imaging. 2020;**2**(2): e200044

[31] Cao Y, Xu Z, Feng J, Jin C, Han X, Wu H, et al. Longitudinal assessment of COVID-19 using a deep learning–Based

quantitative CT pipeline: Illustration of two cases. Radiology: Cardiothoracic Imaging. 2020;**2**(2):e200082

[32] Gaál G, Maga B, Lukács A. Attention U-Net based adversarial architectures for chest X-ray lung segmentation. 2020. arXiv preprint arXiv:2003.10304

[33] Jefferson T, Dooley L, Ferroni E, Al-Ansary LA, van Driel ML, Bawazeer GA, et al. Physical interventions to interrupt or reduce the spread of respiratory viruses. Cochrane Database of Systematic Reviews. 2023. pp. 1-325

[34] Desai AN, Patel P. Stopping the spread of COVID-19. JAMA. 2020; **323**(15):1516-1516

[35] Loey M, Manogaran G, Taha MHN, Khalifa NEM. A hybrid deep transfer learning model with machine learning methods for face mask detection in the era of the COVID-19 pandemic. Measurement. 2020;**167**:108288

[36] Tan C, Sun F, Kong T, Zhang W, Yang C, Liu C. A survey on deep transfer learning. In: Artificial Neural Networks and Machine Learning-ICANN 2018: 27th International Conference on Artificial Neural Networks, Rhodes, Greece, October 4-7, 2018, Proceedings, Part III 27. Springer International Publishing; 2018. pp. 270-279

[37] Wang S, Kang B, Ma J, Zeng X, Xiao M, Guo J. et al. A deep learning algorithm using CT images to screen for corona virus disease (COVID-19). MedRxiv. 2020

[38] Butt C, Gill J, Chun D, Babu BA. Deep learning system to screen coronavirus disease 2019 pneumonia. Applied Intelligence. 2020;**53**(4):4874-4874

[39] Jin C, Chen W, Cao Y, Xu Z, Tan Z, Zhang X, et al. Development and evaluation of an artificial intelligence system for COVID-19 diagnosis. Nature Communications. 2020;**11**(1):5088

[40] Çiçek Ö, Abdulkadir A, Lienkamp SS, Brox T, Ronneberger O. 3D U-Net: Learning dense volumetric segmentation from sparse annotation. In: Medical Image Computing and Computer-Assisted Intervention-MICCAI 2016: 19th International Conference, Athens, Greece, October 17-21, 2016, Proceedings, Part II 19. Springer International Publishing; 2016. pp. 424-432

[41] Isensee F, Petersen J, Klein A, Zimmerer D, Jaeger PF, Kohl S et al. nnU-Net: Self-adapting Framework for U-Net-Based Medical Image Segmentation. In: Bildverarbeitung für die Medizin 2019: Algorithmen–Systeme–Anwendungen. Proceedings des Workshops vom 17. bis 19. März 2019 in Lübeck. Springer Fachmedien Wiesbaden; 2019. pp. 22-22

[42] Ronneberger O, Fischer P, BroxT. U-net: Convolutional networks for biomedical image segmentation. In: Medical Image Computing and Computer-Assisted Intervention-MICCAI 2015: 18th International Conference, Munich, Germany, October 5-9, 2015, Proceedings, Part III 18. Springer International Publishing; 2015. pp. 234-241

[43] Zhou Z, Siddiquee MMR, Tajbakhsh N, Liang J. Unet++: A nested u-net architecture for medical image segmentation. In: Deep Learning in Medical Image Analysis and Multimodal Learning for Clinical Decision Support: 4th International Workshop, DLMIA 2018, and 8th International Workshop, ML-CDS 2018, Held in Conjunction with MICCAI 2018, Granada, Spain, September 20, 2018, Proceedings 4. Springer International Publishing; 2018. pp. 3-11

[44] Milletari F, Navab N, Ahmadi SA. V-net: Fully convolutional neural

networks for volumetric medical image segmentation. In: 2016 4th International Conference on 3D Vision (3DV). Stanford, CA, USA. Oct 2016. pp. 565-571

[45] Oktay O, Schlemper J, Folgoc LL, Lee M, Heinrich M, Misawa K, et al. Attention U-Net: Learning where to look for the pancreas. 2018. arXiv preprint arXiv:1804.03999

[46] Jiang M, Fan X. Retinamask: A face mask detector. 2020. arXiv preprint arXiv:2005.03950

[47] Loey M, Manogaran G, Taha MHN, Khalifa NEM. Fighting against COVID-19: A novel deep learning model based on YOLO-v2 with ResNet-50 for medical face mask detection. Sustainable Cities and Society. 2021;**65**:102600

[48] Howard AG, Zhu M, Chen B, Kalenichenko D, Wang W, Weyand T, et al. MobileNets: Efficient convolutional neural networks for mobile vision applications. 2017

[49] Szegedy C, Vanhoucke V, Ioffe S, Shlens J, Wojna Z. Rethinking the inception architecture for computer vision. In: Proceedings of the IEEE Conference on Computer Vision and Pattern Recognition. 2016. pp. 2818-2826

[50] Simonyan K, Zisserman A. Very deep convolutional networks for large-scale image recognition. arXiv preprint arXiv:1409.1556. 2014

[51] He K, Zhang X, Ren S, Sun J. Deep residual learning for image recognition. In: Proceedings of the IEEE Conference on Computer Vision and Pattern Recognition. Las Vegas, NV, USA. 2016. pp. 770-778

[52] Inamdar M, Mehendale N. Real-time face mask identification using facemasknet deep learning network. 2020. Available from: SSRN 3663305

[53] Chowdary GJ, Punn NS, Sonbhadra SK, Agarwal S. Face mask detection using transfer learning of inceptionv3. In: Big Data Analytics: 8th International Conference, BDA 2020, Sonepat, India, December 15-18, 2020, Proceedings 8. Springer International Publishing; 2020. pp. 81-90

[54] Farooq M, Hafeez A. Covid-resnet: A deep learning framework for screening of covid19 from radiographs. arXiv preprint arXiv:2003.14395. 2020

[55] Hall LO, Paul R, Goldgof DB, Goldgof GM. Finding covid-19 from chest x-rays using deep learning on a small dataset. arXiv preprint arXiv: 2004.02060. 2020

[56] Deng J, Dong W, Socher R, Li LJ, Li K, Fei-Fei L. Imagenet: A large-scale hierarchical image database. In: 2009 IEEE conference on computer vision and pattern recognition. Miami, FL, USA; Jun 2009. pp. 248-255

[57] Karpathy A, Toderici G, Shetty S, Leung T, Sukthankar R, Fei-Fei L. Large-scale video classification with convolutional neural networks. In: Proceedings of the IEEE conference on Computer Vision and Pattern Recognition. Columbus, Ohio, USA; 2014. pp. 1725-1732

[58] Ren S, He K, Girshick R, Sun J. Faster r-cnn: Towards real-time object detection with region proposal networks. Advances in Neural Information Processing Systems. Montreal, Canada: Montreal Convention Center; 2015. p. 28

[59] Long J, Shelhamer E, Darrell T. Fully convolutional networks for semantic segmentation. In: Proceedings of the IEEE Conference on Computer Vision

and Pattern Recognition. Boston, MA, USA; 2015. pp. 3431-3440

[60] Real E, Shlens J, Mazzocchi S, Pan X, Vanhoucke V. Youtube-boundingboxes: A large high-precision human-annotated data set for object detection in video. In: Proceedings of the IEEE Conference on Computer Vision and Pattern Recognition. 2017. pp. 5296-5305

[61] Szegedy C, Liu W, Jia Y, Sermanet P, Reed S, Anguelov D, et al. Going deeper with convolutions. In: Proceedings of the IEEE Conference on Computer Vision and Pattern Recognition. Boston, MA, USA; 2015. pp. 1-9

[62] Wang J, Yang Y, Mao J, Huang Z, Huang C, Xu W. CNN-RNN: A unified framework for multi-label image classification. In: Proceedings of the IEEE Conference on Computer Vision and Pattern Recognition. Las Vegas, NV, USA; 2016. pp. 2285-2294

[63] Yang J, Reed SE, Yang MH, Lee H. Weakly-supervised disentangling with recurrent transformations for 3D view synthesis. In: Advances in Neural Information Processing Systems. Montreal, Quebec, Canada; 2015. pp. 1099-1107

[64] Yim J, Jung H, Yoo B, Choi C, Park D, Kim J. Rotating your face using multi-task deep neural network. In: Proceedings of the IEEE Conference on Computer Vision and Pattern Recognition. Columbus, Ohio, USA; 2015. pp. 676-684

[65] Litjens G, Kooi T, Bejnordi BE, Setio AAA, Ciompi F, Ghafoorian M, et al. A survey on deep learning in medical image analysis. Medical Image Analysis. 2017;**42**:60-88

[66] Kingma DP, Mohamed S, Rezende DJ, Welling M. Semi-supervised learning with deep generative models. Advances in Neural Information Processing Systems. Montreal, Quebec, Canada; 2014. pp. 3581-3589

[67] Laine S, Aila T. Temporal ensembling for semi-supervised learning. arXiv preprint arXiv: 1610.02242. 2016

[68] Rasmus A, Valpola H, Honkala M, Berglund M, Raiko T. Semi-supervised learning with Ladder networks. In: IEEE Conference on Neural Information Processing Systems. Montreal, Quebec, Canada: Neural Information Processing Systems Foundation; 2015. pp. 3546-3554

[69] Weston J, Ratle F, Collobert R. Deep learning via semi-supervised embedding. In: Proceedings of the 25th International Conference on Machine Learning. Helsinki, Finland; Jul 2008. pp. 1168-1175

[70] Lee DH. Pseudo-label: The simple and efficient semi-supervised learning method for deep neural networks. In: Workshop on Challenges in Representation Learning, The 30th International Conference on Machine Learning (ICML 2013), Atlanta, USA. Jun 2013

[71] Berenguer AD, Sahli H, Joukovsky B, Kvasnytsia M, Dirks I, Alioscha-Perez M, et al. Explainable-by-design semi-supervised representation learning for COVID-19 diagnosis from CT imaging. 2020. arXiv preprint arXiv: 2011.11719

[72] Calderon-Ramirez S, Yang S, Moemeni A, Elizondo D, Colreavy-Donnelly S, Chavarria-Estrada LF, et al. Correcting data imbalance for semi-supervised covid-19 detection using x-ray chest images. Applied Soft Computing. 2021;**111**:107692

[73] Ma J, Nie Z, Wang C, Dong G, Zhu Q, He et al. Active contour regularized semi-supervised learning for COVID-19 CT infection segmentation with limited annotations. Physics in Medicine & Biology. 2020;**65**(22):225-234

[74] Paticchio A, Scarlatti T, Mattheakis M, Protopapas P, Brambilla M. Semi-supervised neural networks solve an inverse problem for modeling COVID-19 spread. 2020. arXiv preprint arXiv:2010.05074

[75] Yang D, Xu Z, Li W, Myronenko A, Roth HR, Harmon S, et al. Federated semi-supervised learning for COVID region segmentation in chest CT using multi-national data from China, Italy, Japan. Medical Image Analysis. 2021;**70**: 101992

[76] Zhou J, Jing B, Wang Z, Xin H, Tong H. Soda: Detecting covid-19 in chest x-rays with semi-supervised open set domain adaptation. IEEE/ACM Transactions on Computational Biology and Bioinformatics. 2021;**19**(5): 2605-2612

[77] Abadi M, Agarwal A, Barham P, Brevdo E, Chen Z, Citro C, et al. Tensorflow: Large-scale machine learning on heterogeneous distributed systems. 2016. arXiv preprint arXiv: 1603.04467

[78] Chollet F. et al. Keras. 2015. Available from: https://keras.io

Chapter 10

Smart Healthcare at Home in the Era of IoMT

Qian Qu, Han Sun and Yu Chen

Abstract

Smart Home improves the quality of our life in various aspects such as the convenience of managing our home, efficiency of energy consumption, and secure living environments. Taking advantage of the Internet of Medical Things (IoMT), smart homes in the context of healthcare have attracted a lot of attention to provide a more convenient, easier accessible, and personalized healthcare experience. Leveraging state-of-the-art techniques like Digital Twins (DT), machine learning (ML) algorithms, and human action recognition (HAR), Smart Healthcare at Home (SHAH) not only provides independent healthcare service options and social support but also gives seniors or other individuals who are in need a reliable way for real-time monitoring and safety preservation. This chapter will provide a comprehensive overview of the technical components of a SHAH paradigm, which is based on an architecture that integrates DT, IoMT, and artificial intelligence (AI) technology. The design rationales and key function blocks are illustrated in detail. In addition, taking seniors' safety monitoring as a case study, a prototype of a SHAH system is experimentally investigated, and the performance and design tradeoffs are highlighted. Finally, this chapter also provides an overview of this exciting field's existing challenges and opportunities.

Keywords: internet of medical things (IoMT), smart home, smart healthcare, digital twins, seniors safety monitoring

1. Introduction

Smart Home, or Home Automation, has become one of the trending fields of the Internet of Things (IoT) since 2004. Smart Home refers to a residential space that utilizes advanced information and communication technologies (ICT) and automated systems to enhance comfort, safety, security, and energy efficiency [1]. With smart home devices and systems, homeowners can control and monitor various aspects of their homes, such as lighting, heating, cooling, security systems, and entertainment systems, from anywhere in the world, using their mobile devices or voice assistants. The proposal and development of the Internet of Medical Things (IoMT) bring more elements and functionalities into the landscape of smart homes [2]. Integrated with IoMT, Smart Home can leverage smart sensors to monitor and track residents' vital health data such as blood pressure, heart rate, blood sugar levels, and more. Moreover, personalized health services are provided based on the data collected by the IoMT devices, such as setting up suitable room temperature and lighting schemes to improve the living experience.

A Digital Twin (DT) is a virtual mirror of a physical object or system, such as a building, machine, or city in the digital space, or cyberspace [3]. DT is created by combining data from various sources, such as sensors, cameras, and other IoT devices, and using advanced modeling and simulation tools to create a digital replica of the object or system. In the context of a smart home, a digital twin could be used to create a virtual replica of the home that incorporates real-time data on energy usage, temperature, humidity, and other factors. Moreover, digital twins could then be used to improve the quality of service (QoS) for healthcare in intelligent home systems.

Based on the above rationales, this chapter aims at inspiring more discussions and sparking more new ideas in the digital healthcare community by introducing a scenario of smart healthcare at home (SHAH) that leverages these novel technologies to envision the future of medical services, senior safety, and more exciting application domains.

The rest of this chapter is structured as follows. Section 2 provides necessary background knowledge for readers who are new in this area. Section 3 discusses the design rationals and technical components required to enable smart healthcare at home. Taking the seniors' safety monitoring as a case study, Section 4 illustrates the feasibility of such a framework. Section 5 tries to highlight the major challenges yet to be tackled and the opportunities in the near future. Finally, Section 6 wraps up this chapter with some brief conclusions.

2. Background knowledge

2.1 Internet of Things

Internet of Things (IoT) refers to the network of physical objects or devices that are embedded with sensors, software, and other technologies to collect and exchange data over the Internet. These objects can be anything from simple household appliances such as smart thermostats to more complex devices such as industrial machinery, self-driving cars, and even medical implants. While the potential uses of IoT are virtually limitless, in this chapter we just highlight several most popular areas.

2.1.1 Smart home

A Smart Home is a home equipped with various Internet of Things (IoT) devices that are connected to a network and can be controlled remotely and automatically through a smartphone, tablet, or computer [4]. These devices are designed to make life more convenient, comfortable, and efficient by automating various tasks and functions around the house. Here are some examples of smart home devices:

- Smart thermostats: These devices can be programmed to adjust the temperature of your home automatically based on your preferences and schedule.

- Smart lighting: Smart bulbs and switches can be controlled remotely and programmed to turn on or off at certain times, or in response to other triggers such as motion detection.

- Smart surveillance and locks: These locks can be controlled remotely and allow you to lock or unlock your doors from anywhere, as well as monitor who is coming and going.

- Smart appliances: Many appliances, such as refrigerators, ovens, and washing machines, are now available with IoT connectivity, allowing you to monitor and control them remotely.

- Voice assistants: Devices such as Amazon Echo and Google Home can be used to control your smart home devices using voice commands.

By integrating these devices into a smart home ecosystem, homeowners can automate many tasks and function around the house, making life more convenient and efficient. Li et al. [5] summarized that smart home technology is highly associated with healthcare, energy efficiency, and home security. However, the authors also illustrate challenges such as privacy, security, technology anxiety and negative social influences.

2.1.2 Internet of medical things (IoMT)

IoMT refers to the network of medical devices, wearable sensors, and other healthcare technology that are connected to the Internet and designed to collect, transmit, and analyze patient health data. This data can be used to monitor patient health remotely, diagnose conditions, and deliver more personalized services. Here several examples of IoMT devices are highlighted:

- Wearable fitness trackers that monitor activity levels, heart rate, and sleep patterns.

- Remote monitoring devices that can be used to track vital signs, such as blood pressure, blood glucose levels, and oxygen saturation.

- Medical imaging devices, such as X-ray and MRI machines, that can be connected to the Internet to transmit images and other diagnostic data to healthcare providers.

- Smart pills that contain sensors to monitor medication adherence and provide real-time feedback to healthcare providers.

- Mobile health apps that can be used to monitor and manage chronic conditions such as diabetes, asthma, and heart disease.

IoMT has the potential to revolutionize healthcare by enabling remote monitoring, improving patient outcomes, and reducing costs. For example, IoMT devices can be used to monitor patients with chronic conditions and intervene early if a problem arises, reducing the need for hospitalization and improving quality of life. With the help of state-of-the-art artificial intelligence (AI) techniques, real-time monitoring can be realized using lightweight human action recognition [6].

Additionally, the data collected by IoMT devices can be used to develop more personalized treatments and improve medical research.

2.1.3 Smart grid

A smart grid is an advanced electricity distribution system that uses advanced sensors, communication technologies, and big data to improve the reliability, efficiency, and sustainability of the power grid. The smart grid allows for better integration of renewable energy sources, more efficient distribution of power, and greater control over power usage by both utilities and consumers. By improving the efficiency and reliability of the electric grid, the smart grid can help to decrease energy costs, improve energy security, and reduce greenhouse gas emissions. Additionally, the smart grid can enable the adoption of new energy services, such as electric vehicle charging and home energy management systems. Smart grid technology brings revolutions to energy management from smart homes to smart cities in a "bottom-up approach" [7]. Smart grid not only helps to build a sustainable energy consumption ecosystem for smart homes but also gives new strategies for energy trading on large city-level scales.

2.1.4 Smart city

A smart city is a city that integrates advanced ICT, such as the Internet of Things (IoT), sensors, and data analysis, into its infrastructures, administration, and daily operations to improve the quality of life for its residents, enhance sustainability, and streamline urban services. Smart city initiatives aim to optimize resource management, increase efficiency, and improve communication and connectivity. A smart city consists of various aspects of the daily routines of a modern city operation, including Smart Transportation, Smart Energy, Smart Waste Management, and Smart Public Safety. By using technology to optimize urban services, smart cities can reduce costs and increase efficiency, while also improving the environment and public health.

2.2 Digital twins (DT) and IoMT

The concept of DT was first adopted in the industrial manufacturing domain. As the definition of Physical Object (PO) evolved from industrial artifact into almost every real-world object, DT was soon introduced into the context of healthcare, especially in IoMT. The early examples of DT models benefiting healthcare date back to the maintenance of medical equipment. Until today, the application of this technology in IoMT can be roughly categorized into three major subdomains:

- Medical resource allocation: Leveraging DT models, medical/healthcare service providers can improve the allocation of medical resources including personnel, equipment, medicine, hospital beds, and appointments. The digital models help professionals to better understand the need of patients and predict or estimate future trends. These DT-enabled schemes play a more significant role in crises as was observed during the COVID-19 pandemic as medical institutions may face shortages in various medical resources under extraordinarily heavy pressure. Some hospitals are collaborating with large healthcare enterprises to establish their own DT platforms or systems to provide better QoS for patients [8]. With the help of artificial intelligence (AI) technologies and big data, DT models are powerful to optimize resource allocation and scheduling decisions to enable the entire healthcare service system to be operated more efficiently.

- Digital organs: The process of creating digital organs is different from traditional procedures in manufacturing because most of the data is collected during medical treatment and examinations rather than directly gathered by the sensors [9]. However, based on the massive data about the organs, medical professionals may create DT models and set certain presumptions or scenarios to investigate different hypotheses. This is an effective approach to improve the quality of medical service and reduce the risk of applying new methods to human beings directly. Moreover, combined with certain AI/ML methods, the medical database helps to train various models to obtain more reliable predictions and more accurate diagnoses.

- Digital patient: Digital patient utilizes the real-time collected data from various sensors and/or historical medical records to create a virtual replica of a patient enabling different medical services such as daily examination, remote diagnosis, emergency response [10]. The booming evolution of smart devices especially in IoMT like wearable devices and other light-weight medical sensors brings more opportunities to the development of digital patients. A smart home environment equipped with state-of-the-art IoMT devices allows patients more options to receive personalized medical services. Continuous monitoring and remote medical response systems free certain patients from mandatory hospitalization.

3. Smart healthcare at home (SHAH): An architectural overview

The smart home is one of the major scenarios in which digital patients are enabled. This section presents an architectural overview of a Smart Healthcare at Home (SHAH) system, which integrates DT, IoMT, and AI/ML technologies. By no means the SHAH system covers the entire design space of smart healthcare services in the era of DT, IoMT, and more emerging new technologies, but we hope it provides our readers with some useful insights for further exploration.

3.1 System architecture

Figure 1 illustrates the system architecture of the SHAH framework. This conceptual architecture utilizes DT technology to build an edge-layer virtual healthcare system, which is designed to provide smart healthcare services to the residents living in the smart home. The major components or function blocks in this framework include:

- Body sensors: Body sensors include different types of wearable devices that collect bio-signals such as photoplethysmography (PPG), blood pressure, temperature, heart rate, blood oxygen level. The data collected by these sensors will be transmitted wirelessly to the support unit for modeling and analysis, enabling healthcare professionals to monitor patients remotely and provide timely interventions if necessary.

- Tracking and recognition sensors: This type of sensor mainly refers to smart cameras and motion detectors. With the help of these devices, the system can easily locate the resident(s) and provide more information about the patient. For example, a smart camera can identify different human actions within its field of

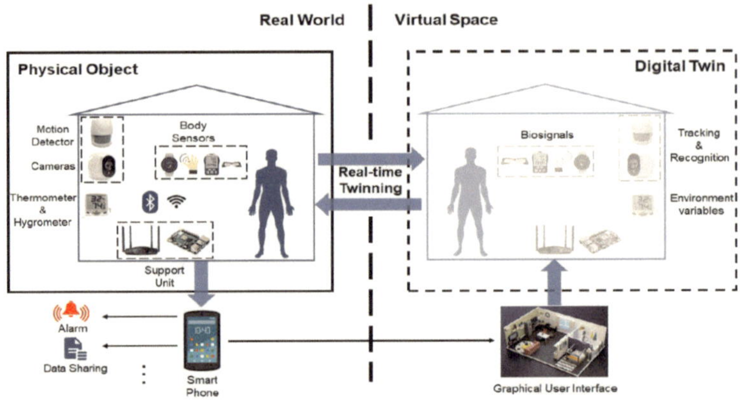

Figure 1.
SHAH system architecture: An overview.

view and the motion detector is triggered if someone enters the room where it was mounted. And by combining this information with other data collected by body sensors, the system is capable of defining certain patterns and giving a timely response if any anomaly is detected.

- Environment sensors: These sensors are responsible for collecting environmental parameters from the facility. Various combinations can be personalized according to usage, such as thermometers, hygrometers, smoke detectors, and water leakage detectors. These environment sensors can be integrated into the centralized smart home system that collects and analyzes data to provide insights and trigger appropriate actions. The collected data can also be shared with healthcare providers to monitor and assess the well-being of individuals remotely or aid in preventive care strategies with the help of AI technologies.

- Support unit: The support unit plays a critical role as the brain of the smart home system. It may consist of a smart gateway, small single-board computers, a personal computer (PC), or even a small home server. The support unit is responsible for collecting all the data from the sensors in the physical world, unifying these parameters in different protocols and sampling rates, providing intelligent healthcare services, and maintaining the DTs in the virtual space.

- Smartphone: Most of the light-weight operations in SHAH can be deployed on a smartphone. The users would have all access to their data and can review the system through a Graphical User Interface (GUI). According to their personal preference, the system can set different alarm patterns and/or data-sharing policies to support customized healthcare services and protect data privacy.

3.2 Technique components

Figure 2 shows the major technique components of SHAH.

- Communication units: SHAH is highly based on the data collected from various sensors, which requires an efficient, accurate, and unified data communication scheme. To address these requirements, technologies such as new-generation

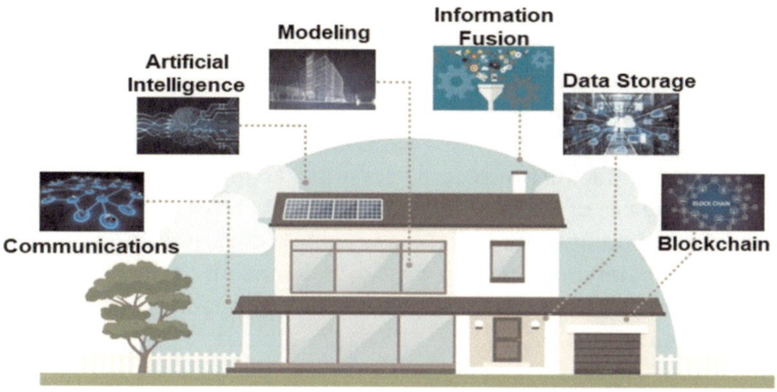

Figure 2.
SHAH technique components.

networks, Bluetooth Low Energy (BLE), Wi-Fi, and various network protocols may be utilized in different scenarios.

- AI: Artificial intelligence plays an important role in SHAH. The massive data collected from sensors relies on AI algorithms to generate different results such as anomaly detection, future prediction, and other insights which would promote healthcare services in this system.

- Modeling and analysis: The DT-based SHAH highly relies on modeling technologies for creating a digital replica inside the virtual space. This digital avatar not only reflects the real-time status of the resident(s) but also gives the user an intuitive vision of the whole living space. Tools like AutoCAD, BIM, Unreal Engine, and 3Dmax are widely used for creating 2D or 3D models of systems.

- Information fusion: Information fusion, also known as data fusion, is the process of integrating and combining data from multiple sources or sensors to obtain more accurate, reliable, and comprehensive information. The goal of information fusion is to leverage the complementary strengths of different data sources, compensate for their individual limitations, and extract valuable insights that would be difficult to obtain from any single source alone.

- Data storage: If we consider the scenario of SHAH as an isolated node that is off the grid, it would be reasonable and practical to store the data on-site and in the system. However, if the system joins the local healthcare network society or even the Metaverse, it involves data sharing and privacy issues. Then, we might introduce distributed data storage which refers to a method of storing data across multiple physical or virtual storage devices or nodes that are geographically distributed. Instead of centralizing data storage in a single location, distributed storage systems distribute data across a network of nodes for improved performance, scalability, fault tolerance, and availability.

- Blockchain and non-fungible tokens (NFT): Security and privacy are paramount in virtual healthcare, where the protection of sensitive medical data during transmission and storage is crucial to prevent unauthorized access and misuse. Blockchain technology holds promise in addressing these concerns by providing a robust solution [11]. Blockchain enables the creation of a secure and tamper-proof ledger for medical records, accessible only to authorized parties. Each transaction or modification to the record is securely recorded on the blockchain, allowing for easy tracking of record access and changes. Another promising technique is the use of NFTs, which possess characteristics such as immutability, traceability, and uniqueness [12]. NFTs can help ensure secure access to medical records while maintaining patient privacy. Ownership of the NFT can be easily transferred to other authorized parties as required, maintaining the integrity and privacy of the records.

4. Seniors' safety monitoring: a case study of SHAH

The entire world is witnessing a fast-growing aging population body, and it is widely observed that there are more and more seniors living alone, either in their homes or in individual rooms of nursing houses. An effective healthcare service system is required to maintain 24/7 real-time monitoring and timely dangerous action recognition. The compelling need for both regular medical consultation and timely emergency assistance inspires us to adopt this topic for our case study.

4.1 Sensor types

There are two types of sensors that are normally used in action recognition: wearable sensors and remote sensors. Wearable sensors are compact and lightweight, making them convenient for real-world settings. They can provide detailed motion data, including acceleration, orientation, and angular velocity information. This type of data helps to recognize specific actions, such as walking, standing, and sitting. However, wearable sensors can be limited in terms of the scope of their coverage. They may not be able to capture specific actions or movements that occur outside of the sensor's range or when the sensor is not worn.

Remote sensors, such as cameras, can capture visual data that complements the motion data provided by wearables. This type of data gives a more comprehensive view of the environment and enables the recognition of complex actions that may not be easily detectable from motion data alone. For example, cameras can capture facial expressions, gestures, and interactions with objects in the environment. However, remote sensors can be limited by lighting conditions, occlusion, and the need for a clear line of sight.

By combining the advantages of both wearable and remote sensors, it is possible to develop a more comprehensive action recognition system that can perform real-time and accurate recognition of a wide range of actions in different settings. In addition, data fusion is a technique used to integrate data from multiple sources to improve the accuracy and reliability of action recognition systems, which can enhance the effectiveness of healthcare and other applications.

4.2 Information fusion

Information fusion is an approach to integrating data from multiple sources to improve the accuracy and comprehensiveness of the information obtained. It is typically divided into three levels: data level, feature level, and decision level [13].

At the data level, raw data from sensors, such as wearable devices and cameras, are combined to represent the phenomenon being monitored thoroughly. This approach can be computationally intensive and requires careful calibration and synchronization of the sensors, but it can provide a more comprehensive understanding of the phenomenon.

At the feature level, features extracted from the raw data are combined to obtain a more informative and complementary representation of the phenomenon. This approach can be more efficient than data-level fusion but requires careful selection and processing of the features to ensure they are informative and complementary.

At the decision level, the decisions or outputs of different classifiers are combined to make a final decision about the phenomenon being monitored. This approach can be used when the various sources of information provide redundant information but are not necessarily complementary. It can also be used to weigh the different sources of data according to their reliability or importance.

In the IoT context, data-level fusion has been preferred due to its ability to integrate data from different sources and provide a more accurate representation of the physical environment. After the data from various sensors are fused at the data level, the resulting data is typically more compact and comprehensive than the raw data from each sensor, as shown in **Figure 3**. Consequently, the amount of data to be transmitted is less than the sample data combination, which is beneficial in conserving network bandwidth and reducing power consumption. In IoT-based action recognition, the fused data can be uploaded to the cloud or processed locally using fewer resources, such as computing power, memory, and energy. By minimizing the resources needed for uploading and processing, the overall system can operate more efficiently and cost-effectively while still achieving high accuracy and real-time performance.

4.3 Data processing methodology

Singular Spectral Analysis (SSA) is a powerful signal-processing technique that has been applied to various domains, including action recognition. In the context of IoT-based action recognition, SSA can be used to implement the skeleton data by combining wearable data [14].

In SSA, time-series data is first embedded into a trajectory matrix, where each row represents a trajectory or a subsequent trajectory of the original data. The trajectory matrix is then decomposed using singular value decomposition (SVD) to separate the data into singular values and corresponding singular vectors. These singular vectors represent the fundamental building blocks of time series. Relevant patterns and features can be extracted by selecting specific singular vectors and their associated singular values.

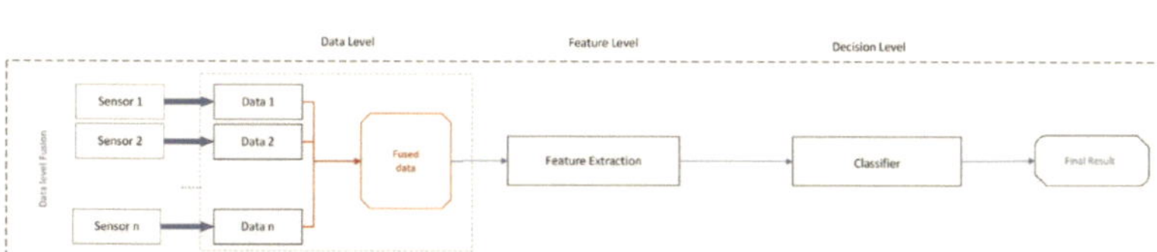

Figure 3.
Data-level information fusion.

The fundamental components obtained by SSA can be interpreted as different patterns of variation in the time series. They capture underlying trends and recurring patterns in the data. These components can be further analyzed and combined to reconstruct the original time series or for various applications such as noise reduction, feature extraction, or forecasting. SSA provides a flexible and practical approach to analyzing time series data and has applications in multiple fields such as finance, climate science, and signal processing.

Figure 4 shows the first ten components sorted by singular value, which is the analysis result of the example of accelerator data in the X-axis when falling. The components are clear enough to represent the trade of initial sensors, which can be used to implement the skeleton data, as it is shown in **Figure 5**.

4.4 Experimental results

Two databases are used in the experimental study. To represent the skeleton data, the NTURGB+D database is chosen. The NTURGB+D (NTU RGB + D) database is a comprehensive benchmark dataset widely used for human action recognition and pose estimation. It comprises a total of 56,880 action samples, recorded from 40 subjects performing 80 distinct actions. Each action has 20 instances, resulting in a diverse set of data for analysis. The database includes RGB videos, depth maps, and skeleton data, providing rich multi-modal information for studying human activities.

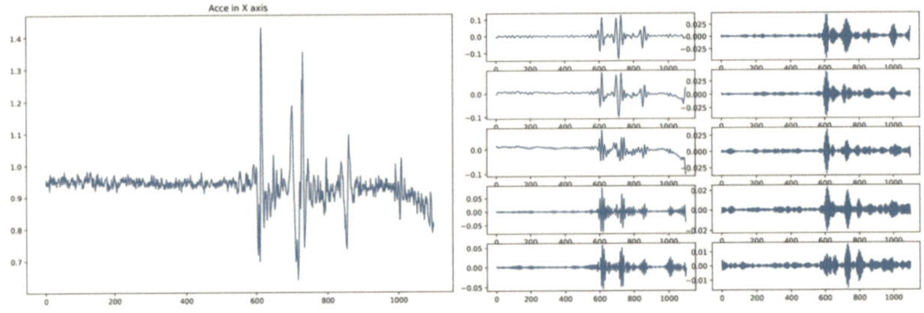

Figure 4.
Accelerator data in X-axis for falling and sequence of the first ten components sorted by singular value.

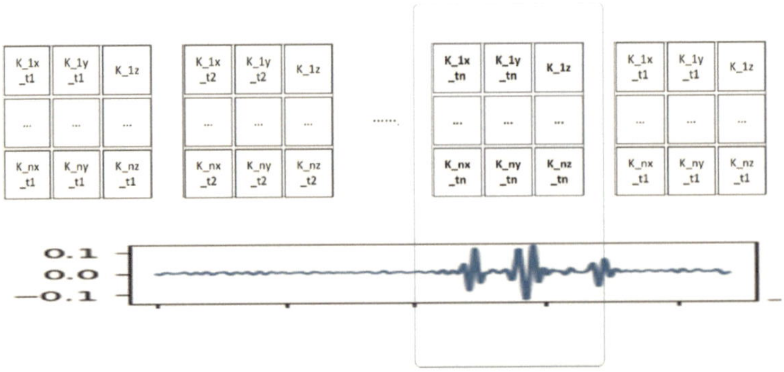

Figure 5.
SSA-based skeleton data implementation.

To represent the initial data, the SCUT-NAA dataset is used. It is a 3D acceleration-based activity dataset that consists of 1278 samples collected from 44 individuals (34 males and 10 females). The data was gathered in naturalistic settings using a single tri-axial accelerometer placed in three different locations: waist belt, pants pocket, and cloth pocket. Each participant was asked to perform ten activities, providing a diverse range of motion data for analysis. To enhance the falling data, the 2015 Fall dataset is added, where data was collected from 32 volunteers specifically focusing on falls. The dataset includes four fall postures: forward, backward, left, and right. Sensors were primarily placed on the chest and thighs to capture both acceleration and angular velocity data during falls. This dataset offers valuable insights into different fall scenarios, enabling researchers to study and develop effective fall detection and prevention algorithms.

After selecting common actions from the above databases and sorting them out, we obtained an experimental dataset with eight actions: Sitting, Walking, Step walking, Jumping, Upstairs, Downstairs, Cycling, and Falling. The total dataset number is 879.

The falling detection result achieved an impressive accuracy rate of 93.82% using the SSA implemented on skeleton data. This outcome demonstrates the effectiveness of the SSA approach in accurately identifying and detecting falls in the dataset. By analyzing the spatiotemporal patterns of skeletal movements, the SSA-based method successfully captured the distinct characteristics associated with falls, leading to highly accurate detection results. The high accuracy rate signifies the potential of SSA-based skeleton data analysis for real-time fall detection systems, which can play a crucial role in ensuring the safety and well-being of individuals, particularly seniors and those at risk of falling. The remarkable performance underscores the value of SSA in enhancing fall detection capabilities and highlights its significance in advancing research and applications in healthcare and eldercare domains.

5. Challenges and opportunities

While the general landscape of DT-enabled smart healthcare at home looks promising and our case study validates its feasibility, there are many open questions yet to be explored. Here, we identify the most critical challenges along with the opportunities in this raising area:

- Security: Although DT technology offers significant advantages for smart healthcare at home, security is still among the top concerns that need to be addressed, such as data integrity, authentication, network security. The integrity of medical/health data is essential and it is mandatory to prevent tampering, manipulation, or unauthorized modifications. Therefore, IoMT-affordable but strong data validation techniques, robust end-to-end data encryption, and secure protocols for data transmission are required. As mentioned in Section 3, a reliable authentication scheme can verify the identity of users accessing DT and their associated healthcare resources. The combination of Blockchain and NFT is a promising approach, but further studies are still needed to address multiple open questions. Network security involves countermeasures to various potential attacks on the smart home network as the system highly relies on wireless-connected devices.

- Privacy: The sensitive medical information asks for strict privacy-preserving protocols. The vital signs, medication history, and data of the resident(s) can be the target of various unauthorized and even malicious agents. Adequate measures such as encryption, access controls, and strict data handling policies should be implemented to ensure the privacy and confidentiality of patient information. Moreover, even authorized data sharing requires traceable, auditable, and transparent operations for users to have full control of their personal medical information. To meet these urgent demands, techniques like dynamic ID changing [15] and reliable time stamp tracking [16] are possible solutions for privacy-preserving.

- Scalability: The scalability problem mainly refers to the efficiency of the system when we expand the smart home into a large-scale network such as a local healthcare community or virtual community in Metaverse. The massive information brought by numerous sensors and devices would have high requirements for the capability of the network.

- Social acceptance: Social acceptance is another important issue as many residents may find it difficult to live with so many sensors, especially cameras. And the sensitive data can cause certain concerns such as misuse or exploitation of these information. Reliability is another problem since the result or decision of the systems is sometimes a matter of life and death.

6. Conclusions and future work

6.1 Conclusions

This chapter envisions the future of digital healthcare services in the IoMT era, a historical time witnessing the proliferation of many new enabling technologies. We discussed the application of Digital Twins, blockchain, and IoMT in the context of the smart home environment. A SHAH framework is introduced as an example of providing real-time monitoring and safety preservation utilizing a combination of different technologies. A preliminary experimental investigation shows the feasibility of using information fusion and statistical analysis to achieve intelligent decision-making for the system. Additionally, several most compelling challenges and open questions are highlighted. We hope this chapter could inspire more discussions in the smart healthcare community and spark new ideas, technical breakthroughs, and novel applications.

6.2 Future work

The preliminary experiment only tested the feasibility of our framework and is limited to a small-scale network. To evaluate the availability of DT-based healthcare services, we need more effort in a large-scale network including the investigation of accuracy and efficiency. Apart from the skeleton recognition and SSA approaches, we will investigate more onsite diagnosis mechanisms and integrate them into SHAH to improve the accuracy of identifying emergent events. Further investigation and studies are required and certain standards need to be established to satisfy local laws and social acceptance.

Abbreviations

AI	artificial intelligence
BLE	Bluetooth low energy
DT	digital twins
GUI	graphic user interface
IoMT	Internet of Medical Things
IoT	Internet of Things
LO	logical object
ML	machine learning
NFT	non-fungible tokens
PC	personal computer
PO	physical object
PPG	photoplethysmography
QoS	quality of service
SSA	singular spectral analysis
SVD	singular value decomposition

Author details

Qian Qu, Han Sun and Yu Chen*
Department of Electrical and Computer Engineering, Binghamton University, Binghamton, New York, USA

*Address all correspondence to: ychen@binghamton.edu

References

[1] Stojkoska BL, Trivodaliev. KV. A review of internet of things for smart home: Challenges and solutions. Journal of Cleaner Production. 2017;**140**:1454-1464

[2] Basatneh R, Najafi B, Armstrong DG. Health sensors, smart home devices, and the internet of medical things: An opportunity for dramatic improvement in care for the lower extremity complications of diabetes. Journal of Diabetes Science and Technology. 2018;**12**(3):577-586

[3] Minerva R, Lee GM, Crespi N. Digital twin in the IoT context: A survey on technical features, scenarios, and architectural models. Proceedings of the IEEE. 2020;**108**(10):1785-1824

[4] Jiang L, Liu D-Y, Yang B. Smart home research. In: Proceedings of 2004 International Conference on Machine Learning and Cybernetics (IEEE Cat. No. 04EX826). Vol. 2. Piscataway, NJ, USA: IEEE; 2004. pp. 659-663

[5] Li W et al. Motivations, barriers and risks of smart home adoption: From systematic literature review to conceptual framework. Energy Research and Social Science. 2021;**80**:102211

[6] Sun H, Chen Y. Real-time elderly monitoring for senior safety by lightweight human action recognition. In: 2022 IEEE 16th International Symposium on Medical Information and Communication Technology (ISMICT). Lincoln, NE, USA: IEEE; 2022. pp. 1-6

[7] Kim H et al. A systematic review of the smart energy conservation system: From smart homes to sustainable smart cities. Renewable and Sustainable Energy Reviews. 2021;**140**:110755

[8] Polyniak K, Matthews J. The johns hopkins hospital launches capacity command center to enhance hospital operations. 2016. Available from: https://www.hopkinsmedicine.org/news/ws/media/releases

[9] Barricelli BR, Casiraghi E, Fogli D. A survey on digital twin: Definitions, characteristics, applications, and design implications. IEEE Acess. 2019;**7**: 167653-167671

[10] Liu Y et al. A novel cloud-based framework for the elderly healthcare services using digital twin. IEEE Access. 2019;**7**:49088-49101

[11] Madine MM et al. Blockchain for giving patients control over their medical records. IEEE Access. 2020;**8**: 193102-193115

[12] Hammi B, Zeadally S, Perez AJ. Non-fungible tokens: A review. IEEE Internet of Things Magazine. 2023;**6**(1):46-50

[13] Zhang Y, Jiang C, Yue B, Wan J, Guizani M. Information fusion for edge intelligence: A survey. Information Fusion. 2022;**81**:171-186

[14] Swart SB, den Otter AR, Lamoth CJC. Singular Spectrum analysis as a data-driven approach to the analysis of motor adaptation time series. Biomedical Signal Processing and Control. 2022;**71**:103068

[15] Algarni A. A survey and classification of security and privacy research in smart healthcare systems. IEEE Access. 2019;**7**:101879-101894

[16] Fan K, Wang S, Ren Y, Yang K, Yan Z, Li H, et al. Blockchain-based secure time protection scheme in IoT. In: IEEE Internet of Things Journal. Vol. 6. No. 3. Piscataway, NJ, USA: IEEE; 2018. pp. 4671-4679